Rewilding European Landscapes

Henrique M. Pereira • Laetitia M. Navarro
Editors

Rewilding European Landscapes

 Springer Open

Editors

Henrique M. Pereira
German Centre for Integrative
Biodiversity Research (iDiv)
Halle-Jena-Leipzig
Leipzig
Germany

Institute of Biology, Martin Luther
University Halle-Wittenberg
Halle (Saale)
Germany

Laetitia M. Navarro
German Centre for Integrative
Biodiversity Research (iDiv)
Halle-Jena-Leipzig
Leipzig
Germany

Institute of Biology, Martin Luther
University Halle-Wittenberg
Halle (Saale)
Germany

ISBN 978-3-319-12038-6   ISBN 978-3-319-12039-3 (eBook)
DOI 10.1007/978-3-319-12039-3
Springer Cham Heidelberg New York Dordrecht London

Library of Congress Control Number: 2014956752

Printed on acid-free paper

Springer is part of Springer Science+Business Media (www.springer.com)

Preface

During the last century humans have dramatically accelerated alterations and loss of biodiversity worldwide. Changes to our planet's ecosystems by Humans go back tens of thousands of years, but what happened in the last couple hundred years has no precedent in the history of our species. We took habitat change, overexploitation, biotic homogenization, and pollution to a new level. We even started to change the Earth's climate, a feat perhaps never achieved by any other single species. Today, with a human population of already over 7 billion, about 40 % of the world's forests and other natural ice-free habitats have been converted to cropland and pasture, we have appropriated 15 % of global terrestrial net primary production, and species extinction rates are 100 times greater than the average extinction rate for the Cenozoic fossil record.

But this book is not about this ecological disaster. Instead, this book is about a new conservation strategy that brings hope for restoring some of the lost biodiversity and ecosystem functions. This book is about rewilding. Rewilding is the passive management of ecological succession with the goal of restoring natural ecosystem processes and reducing the human control of landscapes. The opportunity for large-scale rewilding in Europe has been developing over the last few decades through the process of land abandonment, particularly farmland abandonment. Some projections estimate that between 2000 and 2030 as much as 20 million ha may be released from agricultural use in Europe, an area twice the size of Portugal.

Farmland abandonment has been raising much concern in the scientific and policy communities. There are grave social implications, with rural exodus, the aging of rural communities, and the decrease of basic public services for the populations that stay in the abandoned areas. But, there have also been ecological concerns. Many of the abandoned landscapes where associated with low intensity farming systems or semi-natural grasslands that host high biodiversity. The fear has been that, without the maintenance of those systems, much biodiversity and ecosystem services will be lost. Therefore measures such as agro-environmental subsidies and support to least favored areas have been implemented under the Common Agricultural Policy.

Rewilding is an alternative approach to the management of these systems. In rewilding the goal is not to maintain some habitats in a static snapshot of the past. Rewilding embraces change in ecosystems and aims at restoring ecological dynamics that run their course independent of human intervention, including disturbances such as fire and diseases. Rewilding does not target to conserve species occurrences in a given set of sites, but instead aims at allowing large scale processes involved in population and community dynamics to reestablish themselves. These processes have particularly been hampered for the European megafauna in the last few centuries and therefore rewilding also promotes wildlife comeback. Ultimately, rewilding is better defined by the strategies that aim at allowing natural processes to regain dominance in landscapes rather than by some fixed set of end goals for how the landscape and biodiversity should look like. Rewilding is about having large populations of large herbivores, with space for movements tracking the seasons and interannual variation in resources, but also about having predators that keep these herbivores in check. It is about having healthy necrophagous bird communities living in an unpredictable environment and providing us ecosystem services instead of artificially feeding those communities in vulture restaurants. Rewilding is also a process that is continuously on-going: ecosystems progressively become more natural over time and although effects are already being seen and will be seen in the policy relevant time scale of a few decades, it will continue to evolve during a much longer time frame.

However, we should caution that the rewilding we propose is different from other rewilding approaches, and we shall refer it as ecological rewilding to make the distinction clear. Europe is a densely populated continent that has been deeply modified over the last millennia. By the height of the Roman Empire, much of the European forest had already been cleared, and today Europe is crossed with roads, punctuated by villages, and managed in large regions by foresters and farmers. Ecological conditions today are very different from what they were at the end of the Pleistocene. Europe is now much warmer (and becoming warmer) and wetter. Soils have been highly modified by agriculture, and almost no place is more than a few kilometers away from human settlements. In such a context, something akin to Pleistocene rewilding, where the goal is to go back to some historic baseline, for instance by bringing back the ecological equivalents of the mammoth and the cave lion, can only perhaps be implemented in a zoo fashion: with fences separating people from animals. Instead, in ecological rewilding we recognize that in Europe, as in many other parts of the world, we manage a complex socio-ecological system where humans are an integral component of our landscapes. Historical baselines are useful to inspire our management strategies in ecological rewilding as they help us understand how ecosystems function in the relative absence of human intervention, but they are not the goals of ecological rewilding. Ecological rewilding is about using that historical information and the best available ecological knowledge to design conservation strategies, sometimes involving at the beginning active management. It aims at restoring ecosystems where human control of ecological processes is much reduced, wildlife strives, and non-extractive ecosystem services such as carbon sequestration and recreation are provided. Ecological rewilding is also not centered in species reintroductions. Reintroductions may be advantageous in many rewilding projects, but they need to be assessed case by case, and they need to be incorporated in a wider strategy to restore natural disturbance regimes and natural succession.

This book brings together contributions from thirty authors across nine different countries to discuss rewilding of abandoned landscapes in Europe. We bring together scientists and practitioners, as rewilding is at the interface of science and society, and we target as well an audience of thinkers and doers. We do not shy away from controversy or critical views, and some of the chapters present different perspectives from the book editors' on how to manage biodiversity and farmland abandonment. The book has a clear European focus, but the approaches, results and discussion is certainly relevant worldwide, as abandonment is at least a local phenomenon in all parts of the world.

The first part of the book aims at developing the basis of a theory of rewilding. Chapter 1 by Pereira and Navarro lays out the basic ideas for the book, and is a reprint of our original paper on this topic in the journal *Ecosystems*. The chapter questions traditional paradigms of managers and scientists on European landscapes, such as the sustainability of traditional farming practices, the quality of life of rural populations, and the efforts to maintain museum landscapes. The authors argue that farmland abandonment is inexorable and that concerns with negative impacts on biodiversity may be unfounded as the relationship between species diversity and land-use has not been examined at large scales. The authors identify winner and looser species of rewilding and argue that rewilding can drive important changes in community composition. The chapter examines the benefits of rewilding based on an ecosystem service framework, and concludes by identifying some challenges associated with rewilding such as conflicts with wildlife and limits to ecological resilience.

Chapter 2, by Ceauşu and colleagues, discusses the concept of wilderness by analyzing how it can be mapped in space. The chapter briefly reviews the historical concepts of wilderness and wilderness metrics, discussing the subjectivity of wilderness perception. The authors map wilderness areas in Europe using four concepts: (1) remoteness from roads and settlements; (2) absence of light pollution; (3) distance to potential natural vegetation; and (4) proportion of primary productivity harvested by humans. They show that wilderness areas are concentrated in high latitudes and in mountainous regions and that they overlap with areas of high megafauna species richness. Surprisingly, they find even higher values of wilderness in areas of Natura 2000 than in the nationally protected areas. Finally, the authors argue that farmland abandonment will occur in areas of intermediate wilderness values, releasing additional areas for wild ecosystems.

Chapter 3, by Cerqueira and colleagues, examines the consequences of rewilding for ecosystem services. Using a spatial map of the distribution of the different types of ecosystem services in Europe, the authors investigate the spatial distribution of indicators of ecosystem services in Europe before analyzing the spatial overlap between wilderness and areas delivering provisioning, regulating and cultural services. They then proceed to a quantitative analysis of the average supply of services between areas of wilderness, areas under agricultural use, and areas projected to be abandoned. Their analysis suggests that wilderness areas provide important recreation services, carbon sequestration, nitrogen retention. Furthermore, the agricultural productivity of areas to be abandoned is much lower than of areas that are currently used for agriculture, suggesting that the impact of abandonment in food production in Europe is limited. The chapter concludes with an analysis of the economic benefits of the cultural services associated with rewilding.

The second part of the book discusses the consequences of rewilding for biodiversity. Chapter 4, by Boitani and Linell, reviews the wildlife comeback in Europe and suggests a new conservation strategy for large carnivores. After the near extinction of wolves, bears, and Eurasian lynxes, in the first half of the twentieth century, these species have experienced a continent wide recovery. The chapter examines the causes for this recovery, which are a complex interaction between changes in public opinion towards wildlife, species protection including by better hunting management, reintroductions, and habitat recovery after abandonment. Next, the authors examine how to move from conservation strategies targeted at averting extinction to strategies targeted at sustaining recovery, and question how far that recovery should go. They present a critical view of rewilding, and instead argue for a strategy aiming at orientated coexistence between wildlife and humans by avoiding human-wildlife conflicts.

Chapter 5, by Cortés-Avizanda and colleagues, discusses the conservation of avian scavengers in the context of farmland abandonment and rewilding. Vultures originally depended on the availability of wild ungulate carcasses. With the progressive domestication of Europe, vultures switched to livestock as their main food source. However, in the last few decades, poisoning and the decrease in availability of livestock carcasses due to sanitary regulations, lead to a dramatic decline of these species. In response, a conservation strategy was developed based on supplementary feeding at "vulture restaurants". This was largely successful for the targeted species, but changed the spatial-temporal nature of food resources, shifting the balance in scavenger communities towards those target species. The authors examine alternatives to vulture restaurants opened by rewilding, such as the increase in wild ungulate populations, and consider complementary approaches, such as the promotion of extensive agro-grazing practices. The chapter concludes with a review of the ecosystem services provided by vultures.

Chapter 6, by Merckx, looks at two taxonomic groups that are not often discussed in the context of rewilding: moths and butterflies. As other aspects of biodiversity, butterflies and moths have been in decline in Europe, and although agricultural intensification is behind decreases in both groups, farmland abandonment has also been involved in butterfly declines. The nocturnal and endothermic behavior of moths and their association to forest habitats are a possible explanation for this difference. Merckx then presents a study of macro-moth diversity in a landscape undergoing abandonment in the Peneda mountain range. He finds that across a range of spatial scales, forest-dominated landscapes have higher species diversity than shrub-dominated or meadow dominated landscapes. Furthermore, the diversity of closed-biotope species increases faster with spatial scale than the diversity of open-biotope species, suggesting some of the positive effects of rewilding can only be identified at larger scales.

Chapter 7, by Benayas and Bullock, discusses the challenges of restoring forests on agricultural land where tree recruitment is limited either by soil degradation or limited seed dispersal, and frame that discussion in the "land sharing" versus "land sparing" debate. They advocate a proactive approach closer to land sharing based on the strategic revegetation of farmed fields. The idea is to plant woodland islets and hedgerows in a tiny fraction (<1 %) of the target area. This will have an immediate positive impact on biodiversity, by creating a more heterogeneous landscape and providing habitat for forest dependent species. In case the landscape becomes abandoned, dispersal of seeds by wind and animals from the woodland islets and hedgerows will allow for a faster revegetation. They conclude by examining options for forest restoration in a land separation (i.e. lands sparing) context, comparing passive rewilding with forest plantations.

Chapter 8, by Navarro and colleagues, examines the role of disturbance in rewilding. Many species are associated with open habitats created by disturbances. The chapter discusses two important disturbances in pre-Neolithic landscapes: fire and herbivores. Next, it examines the relationship between a disturbance regime and biodiversity. Intermediate levels of disturbance are often associated with higher species diversity, as both early colonizers and more late-successional species can co-occur. Therefore it is important to maintain, at large scales, a heterogeneous landscape where disturbances have a stochastic distribution. The chapter concludes with a discussion of how to reestablish natural regimes of these disturbances. The authors briefly examine which situations are more amenable for natural recolonization of ungulates and in which situations active reintroduction should be preferred. Prescribed burning is discussed as a technique to help reintroducing more natural fire dynamics in ecosystems.

The third and final part of the book examines examples of how rewilding can be put in practice. Chapter 9, by Wouter and colleagues, presents the Rewilding Europe initiative, which aims at restoring wilderness on 1million ha of abandoned farmland. The Rewilding Europe initiative is built around three pillars: the conservation actions targeted at rewilding, communicating rewilding to the public, and developing local enterprises. The chapter presents the 5 locations across Europe where the Rewilding Europe Foundation teamed up with local NGOs to implement the rewilding of an area. It discusses the challenges faced in each location and the different local contexts and goals for rewilding.

Ultimately, for rewilding to be successful it must also help the local economies. Chapter 10, by Jobse and colleagues, further elaborates on on-going efforts to support the development of enterprises associated with rewilding, through an education program. The chapter presents an Erasmus Intensive Program, which is training a new generation of wilderness entrepreneurs. The chapter discusses the design of the curriculum, the competences that the program aims to develop, and the learning environment that is promoted. This discussion is based on the experience of the first year of the program, where students visited and developed activities in the Rewilding Europe site of Western Iberia.

The book concludes with a chapter on laying the foundations for a European policy for rewilding. Chapter 11, by Navarro and Pereira, starts by providing a short historical perspective on nature conservation. It discusses current policies with relevance for nature conservation in the EU, including the Nationally Designated Protected Areas, the Habitats and the Birds Directives, and the second pillar of the Common Agricultural Policy. The authors argue that wilderness protection has received little attention in Europe, although there are encouraging recent developments. The chapter discusses on how to build on those developments to widen the scope of policies targeted towards wilderness protection and rewilding. Finally, the authors argue that rewilding can play a major role in achieving some of the 2020 targets of the Convention on Biological Diversity and the EU Biodiversity Strategy.

We hope that this book inspires more research on rewilding and more practitioners to push the boundaries of conservation strategies. Nature is resilient but we need to learn how to better support that resilience. We conclude with a personal note on that resilience and on its limits. Henrique has been doing fieldwork in Peneda-Gerês National Park in Northern Portugal for over 25 years. Back in the early 1990's this was an area where observing a wild boar or a roe deer was a rare event. A neighbor of Henrique's in Castro Laboreiro shared once that his grandmother, whom lived through most of the twentieth century, had never seen a wild boar. Now, things have changed. Today you can hardly spent more than a few weeks in the region without seeing a wild boar or a roe deer. A combination of farmland abandonment, reduced livestock grazing, and species protection with limited hunting was probably behind this wildlife comeback. Yet, there was a species that would not have come back by itself as it went locally extinct in the beginning of the twentieth century: the Iberian Ibex (*Capra pyrenaica*). The species has been successfully reintroduced in the Spanish side of the Gerês mountain range in the 1990s and in the Spanish side of the Laboreiro mountains in the late 2000s. They now can be seen in both regions in the Peneda-Gerês National Park. And what an amazing sight it is to watch these agile animals climbing the large granite domes!

We wish to dedicate this book to all the managers, scientists and last but not the least, the local communities, which made this wildlife return possible.

Henrique M. Pereira
Laetitia M. Navarro

Contents

Part III   Rewillding in Practice

Contributors

Luigi Boitani Department of Biology and Biotechnologies, University of Rome "La Sapienza", Rome, Italy

Lluis Brotons European Bird Census Council (EBCC), Centre de Recerca Ecològica i Aplicacions Forestals (CREAF), Centre Tecnològic Forestal de Catalunya (CEMFOR—CTFC), Solsona, Spain

James M. Bullock Centre for Ecology and Hydrology, Wallingford, Oxfordshire, UK

Steve Carver Wildland Research Institute, School of Geography, University of Leeds, Leeds, UK

Silvia Ceaușu German Centre for Integrative Biodiversity Research (iDiv) Halle-Jena-Leipzig, Leipzig, Germany

Institute of Biology, Martin Luther University Halle-Wittenberg, Halle (Saale), Germany

Centro de Biologia Ambiental, Faculdade de Ciências da Universidade de Lisboa, Lisboa, Portugal

Yvonne Cerqueira Centro de Biologia Ambiental, Faculdade de Ciências da Universidade de Lisboa, Lisboa, Portugal

Centro de Investigação em Biodiversidade e Recursos Genéticos (CIBIO), Departamento de Biologia, Faculdade de Ciências da Universidade do Porto, Porto, Portugal

Ainara Cortés-Avizanda Centro de Biologia Ambiental Faculdade de Ciências da Universidade de Lisboa, Lisboa, Portugal

Department of Conservation Biology, Estación Biológica de Doñana, CSIC, Sevilla, Spain

José A. Donázar Department of Conservation Biology, Estación Biológica de Doñana, CSIC, Sevilla, Spain

Wouter Helmer Rewilding Europe, Nijmegen, The Netherlands

Franz Hölker Leibniz Institute of Freshwater Ecology and Inland Fisheries, Berlin, Germany

Judith C. Jobse Van Hall Larenstein University of Applied Sciences, Velp, The Netherlands

Jed O. Kaplan Institute of Earth Surface Dynamics University of Lausanne Geopolis, Lausanne, Switzerland

Helga U. Kuechly Leibniz Institute of Freshwater Ecology and Inland Fisheries, Berlin, Germany

John D. C. Linnell Norwegian Institute for Nature Research, Trondheim, Norway

Joachim Maes European Commission, Joint Research Centre, Sustainability Assessment Unit, Ispra, VA, Italy

Cristina Marta-Pedroso IN+, Center for Innovation, Technology and Policy Research, Environment and Energy Scientific Area, Instituto Superior Técnico, University of Lisbon, Lisboa, Portugal

Thomas Merckx Behavioural Ecology and Conservation Group, Biodiversity Research Centre, Earth and Life Institute, Université catholique de Louvain (UCL), Louvain-la-Neuve, Belgium

Centro de Biologia Ambiental, Faculdade de Ciências da, Universidade de Lisboa, Lisboa, Portugal

Laetitia M. Navarro German Centre for Integrative Biodiversity Research (iDiv) Halle-Jena-Leipzig, Leipzig, Germany

Institute of Biology, Martin Luther University Halle-Wittenberg, Halle (Saale), Germany

Centro de Biologia Ambiental, Faculdade de Ciências da Universidade de Lisboa, Lisboa, Portugal

Henrique M. Pereira German Centre for Integrative Biodiversity Research (iDiv) Halle-Jena-Leipzig, Leipzig, Germany

Institute of Biology, Martin Luther University Halle-Wittenberg, Halle (Saale), Germany

Centro de Biologia Ambiental Faculdade de Ciências da Universidade de Lisboa, Lisboa, Portugal

João Pradinho Honrado Centro de Investigação em Biodiversidade e Recursos Genéticos (CIBIO), Departamento de Biologia, Faculdade de Ciências da Universidade do Porto, Porto, Portugal

Vânia Proença IN+, Center for Innovation, Technology and Policy Research, Environment and Energy Scientific Area, DEM, Instituto Superior Técnico, University of Lisbon, Lisboa, Portugal

José María Rey Benayas Departamento de Ciencias de la Vida—UD Ecología, Universidad de Alcalá, Alcalá de Henares, Spain

Deli Saavedra Rewilding Europe, Nijmegen, The Netherlands

Judith Santegoets Van Hall Larenstein University of Applied Sciences, Velp, The Netherlands

Frans Schepers Rewilding Europe, Nijmegen, The Netherlands

Derk Jan Stobbelaar Van Hall Larenstein University of Applied Sciences, Velp, The Netherlands

Magnus Sylvén Rewilding Europe, Nijmegen, The Netherlands

Peter H. Verburg Institute for Environmental Studies (IVM), VU University Amsterdam, Amsterdam, HV, The Netherlands

Loes Witteveen Van Hall Larenstein University of Applied Sciences, Wageningen, The Netherlands

List of Figures

List of Tables

Part I
The Theory of Rewilding

Chapter 1
Rewilding Abandoned Landscapes in Europe

Laetitia M. Navarro and Henrique M. Pereira

Abstract For millennia, mankind has shaped landscapes, particularly through agriculture. In Europe, the age-old interaction between humans and ecosystems strongly influenced the cultural heritage. Yet European farmland is now being abandoned, especially in remote areas. The loss of the traditional agricultural landscapes and its consequences for biodiversity and ecosystem services is generating concerns in both the scientific community and the public. Here we ask to what extent farmland abandonment can be considered as an opportunity for rewilding ecosystems. We analyze the perceptions of traditional agriculture in Europe and their influence in land management policies. We argue that, contrary to the common perception, traditional agriculture practices were not environmentally friendly and that the standards of living of rural populations were low. We suggest that current policies to maintain extensive farming landscapes underestimate the human labor needed to sustain these landscapes and the recent and future dynamics of the socio-economic drivers behind abandonment. We examine the potential benefits for ecosystems and people from rewilding. We identify species that could benefit from land abandonment and forest regeneration and the ecosystem services that could be provided such as carbon sequestration and recreation. Finally, we discuss the challenges associated with rewilding, including the need to maintain open areas, the fire risks, and the conflicts between people and wildlife. Despite these challenges, we argue that rewilding should be recognized by policy-makers as one of the possible land management options in Europe, particularly on marginal areas.

H. M. Pereira (✉) · L. M. Navarro
German Centre for Integrative Biodiversity Research (iDiv) Halle-Jena-Leipzig, Deutscher Platz 5e, 04103 Leipzig, Germany
e-mail: hpereira@idiv.de

Institute of Biology, Martin Luther University Halle-Wittenberg, Am Kirchtor 1, 06108 Halle (Saale), Germany

Centro de Biologia Ambiental, Faculdade de Ciências da Universidade de Lisboa, Campo Grande, 1749-016 Lisboa, Portugal

L. M. Navarro
e-mail: laetitia.navarro@idiv.de

H. M. Pereira, L. M. Navarro (eds.), *Rewilding European Landscapes,*
DOI 10.1007/978-3-319-12039-3_1, © The Author(s) 2015

Keywords Farmland abandonment · Land-use change · Passive management · Ecosystem services · Land sharing · Land sparing

1.1 Introduction

Deforestation and the loss of natural habitats remain major global concerns. None-theless, although scenarios for the next decades project the continuation of these dynamics in tropical ecosystems, the projections made for much of the Northern Hemisphere are quite the opposite (Pereira et al. 2010). In fact, most deforestation in Europe occurred before the industrial revolution (Kaplan et al. 2009), and the amount of forests and scrubland is now increasing following the land abandonment that began in the mid-twentieth century (FAO 2011), a trend that is expected to continue over the next few decades (van Vuuren et al. 2006). Natural vegetation recovery is a complex process that occurs during the progressive alleviation of agricultural use (Hobbs and Cramer 2007; Stoate et al. 2009). This reduction in land-use intensity, including abandonment at the extreme, is, at the local scale, explained by a combination of socio-ecological drivers (MacDonald et al. 2000; Rey Benayas et al. 2007) such as low productivity and aging of the population. These factors interact between them and with the ecological dynamics of succession, creating positive feedback loops, which increase the irreversibility of farmland abandonment in marginal areas, and reduce the effectiveness of subsidies awarded to farmers to halt abandonment (Figueiredo and Pereira 2011; Gellrich et al. 2007). In Europe, there has been a decline of 17 % of the rural population since 1961 (FAOSTAT 2010). Some parishes of Mediterranean mountain areas have lost more than half of their population in a similar period (Gortázar et al. 2000; Pereira et al. 2005). At the regional scale, the current farmland contraction is best explained by an increase in agricultural productivity and the slowing of population growth in Europe (Keenleyside and Tucker 2010).

Landowners and managers facing increased agricultural market competition have resorted mostly to one of three active management strategies (Fig. 1.1): intensification, extensification, and afforestation. Intensification is often chosen on the most productive soils and where good conditions exist for mechanization (Pinto-Correia and Mascarenhas 1999). Extensification consists of obtaining higher productivity by expanding the area of the farm through land consolidation or in developing multiple uses of the land. This has happened in the Montado and Dehesa areas of Portugal and Spain, an agroforestry system that integrates animal production, cork harvesting and cereal cultivation, while hosting high biodiversity and providing recreational and aesthetical benefits (Bugalho et al. 2011). Finally, in some areas with poor farmland soils, the option has been to plant forests, often of fast growing species (Young et al. 2005).

In this article, we discuss a fourth option: rewilding abandoned landscapes, by assisting natural regeneration of forests and other natural habitats through passive management approaches. Rewilding has seldom been considered as a land management policy, as often it faces resistance from both the public (Enserink and Vogel 2006; Bauer et al. 2009) and the scientific communities (Conti and

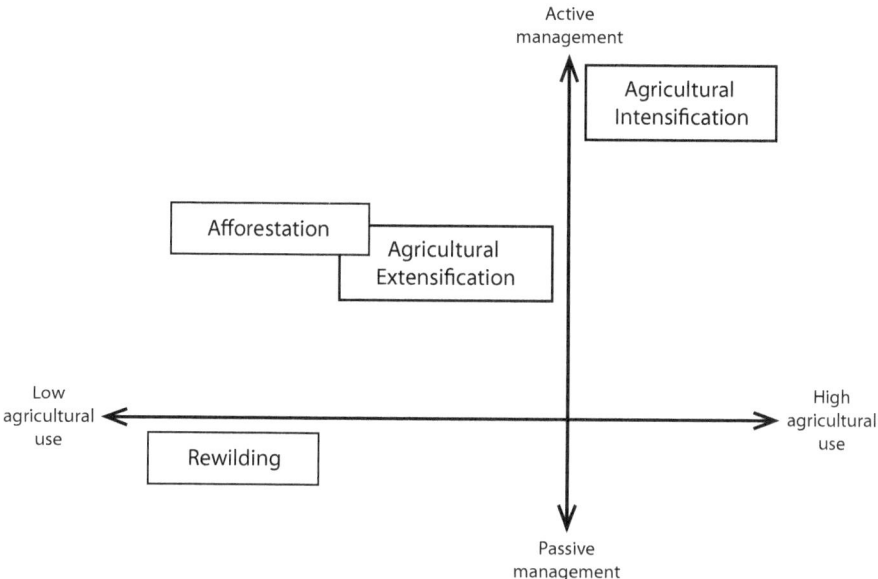

Fig. 1.1 Landscape management strategies plotted against agricultural use intensity and level of management (from active to passive): agricultural intensification, agricultural extensification, afforestation, and rewilding

Fagarazzi 2005; Moreira and Russo 2007). Arguments against rewilding include the loss of the traditional agricultural landscape and negative impacts on biodiversity and ecosystem services (for example, Conti and Fagarazzi 2005). This situation has given rise to a pattern of double standards: developing countries are asked to halt deforestation while some developed countries are actively fighting forest regeneration on their own land (Meijaard and Sheil 2011).

Here, we critically examine some of the arguments used in support of the maintenance of the traditional landscapes and contrast those arguments with the potential benefits for ecosystem services and biodiversity that could accrue from rewilding. We conclude with an analysis of the main challenges associated with rewilding abandoned landscapes.

1.2   European Landscapes: Examining the Paradigms

The cultural importance of traditional agriculture landscapes has been widely recognized in Europe and the world. As of 2011, 76 of the 936 UNESCO world heritage sites are in the "cultural landscapes" category (http://whc.unesco.org), and 29 of those because of traditional or symbolical agricultural practices. Examples include the "Causses and Cevennes Mediterranean agro-pastoral cultural landscape" in France or the "Mont Perdu" in the Pyrénées. As much as 15–25 % of the European farmland can be classified as High Nature Value farmland (EEA 2004).

Of the 231 habitat types listed in the European Habitats Directive, 41 are associated with low-intensity agricultural management, including semi-natural grasslands and hay meadows (Halada et al. 2011).

This has lead to a generalized push towards policies embracing the protection of extensive farming systems with the dual-role of protecting biodiversity and ecosystem services. Here we argue that not all socio-ecological aspects of the maintenance of these landscapes have been taken into account because our perceptions of these landscapes have been biased by our own cultural experiences. We question three ideas associated with current policies: (1) the idea that traditional agriculture practices were environmentally friendly; (2) the idea that traditional rural populations lived well; (3) the idea that traditional landscapes can be kept despite the context of recent rural exodus and future socioeconomic trends.

Were Traditional Agricultural Practices Environmentally Friendly?

In Europe, pre-Neolithic Holocene landscapes can most likely be described as a mosaic of old-growth forest, scrubland, and grasslands, maintained by the grazing of large herbivores and by fire (Svenning 2002; Vera 2000, Vera 2009), although the relative amount of open area is debated (for example, Hodder et al. 2009). Later on, and much before the onset of modern agriculture, European inhabitants destroyed most of Europe's forests on usable land. Europe is now the continent with the least original forest cover (Kaplan et al. 2009).

The process of forest clearing might be as old as human's making of tools (Williams 2000). It started in the Neolithic with the use of fire to open areas for grazing and hunting (Pereira et al. 2012). Forest loss was accelerated during Antiquity, when the rise of classical civilizations led to large-scale deforestation (Williams 2000; Kaplan et al. 2009). After a brief interruption caused by the breakdown of the Roman society, the deforestation trend continued in the Middle Ages (interrupted only by the Black Death), with an estimated loss of 50–70% of the European forest during this period.

Hence humans amplified the disturbance regime of European ecosystems and expanded the open area considerably (Pereira et al. 2012, see Chap. 8), creating and maintaining "traditional" landscapes such as the alpine grasslands (Laiolo et al. 2004), and the agro-silvo-pastoral systems of Mediterranean regions (Blondel 2006). These extensive farming systems have higher species diversity than intensive farming systems (Batáry et al. 2012; Tscharntke et al. 2005), and, at the local scale, often have higher species diversity than non-managed ecosystems and natural forests (Blondel 2006; Höchtl et al. 2005; Lindborg et al. 2008). Therefore, it has been suggested that biodiversity peaks for low levels of land use associated with these extensive farming systems (Fig. 1.2), following the intermediate disturbance principle (Wilkinson 1999).

This pattern has been used as an argument to maintain the active management of extensive farmland and halt ecological succession. However at regional scales, this relationship is likely to exhibit a different pattern (Fig. 1.2). The habitat turnover of wild landscapes can be a mosaic of closed forest and open areas, which should

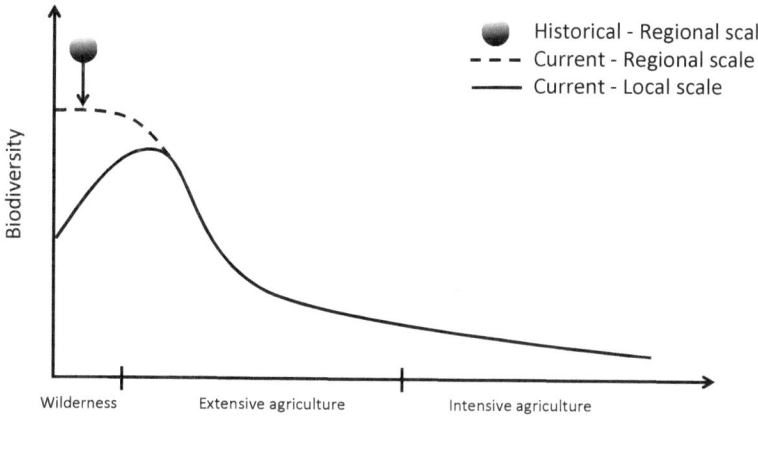

Fig. 1.2 Conceptual representation of the response of current species diversity to land-use intensity at the local and regional scales, and of the hypothetical regional response if Holocene extinctions had not occurred. The response at the local scale is adapted from EEA (2004), whereas the current and historical responses at the regional scale are discussed in the text

accommodate many of the species that can usually be found in extensive farmland habitats. In the early Holocene, the regional diversity of wild landscapes would have been even higher (Fig. 1.2). Several species have now disappeared due to the expansion of human activities, including the auroch (*Bos primeginius*), the Tarpan (*Equus ferus ferus*), or became extinct in most of their former ranges (for example, wisent, *Bison bonasus*).

Deforestation also had important impacts on ecosystem services. In the Mediterranean basin, deforestation is thought to have caused desiccation and soil erosion (McNeely 1994; Blondel 2006). In the Middle Ages, timber shortage is likely to have played a role on the impulse to conquer new territories (Farrell et al. 2000). To build naval fleets, countries such as Portugal and Spain had to resort to importing wood from colonies from the sixteenth century on (Devy-Vareta and Alves 2007). By the end of the nineteenth century, the dimension of the erosion problems in mountain slopes and associated silting in rivers and floods downstream led to large state sponsored afforestation programs in Portugal and Spain.

Did Traditional Rural Populations Live Well?

For centuries, populations inhabiting marginal agricultural areas organized their lives in a self-sufficient manner (Blondel 2006). The industrial revolution and the globalization of the food and labor markets brought many of these regions to an economic disadvantage with urban and peri-urban areas: increasing wages associated with economic growth and the low food prices in global markets rendered the low-productivity farmland uncompetitive.

Nowadays, marginal agricultural areas throughout the globe are classified as "poverty traps" where households suffer from scarcity of resources, low return on investment, lack of opportunities, and reduced social services (Conti and Fagarazzi 2005; Ruben and Pender 2004). For example, in mountains of Southern Europe, rural populations are constrained by the low productivity of small-scale parcels and the limited opportunities for mechanization and intensification (MacDonald et al. 2000). On average, across European mountain areas, the income per hectare is about 40 % lower than in other, non-disadvantaged, areas (809 €/ha vs. 1370 €/ha in EC 2009). The young have limited access to education and employment while the elders experience isolation and difficulties to access services (EC 2008a). This results in out-migration and aging of the population, leading to an inverted population pyramid. This rural exodus is driven by a "circle of decline" where low population density limits business creation, causing fewer jobs and more out-migrations which, in turn, accentuates the decrease in population density (EC 2008a).

Rural populations still value the quality of their environment and its scenic beauty (Bell et al. 2009; Pereira et al. 2005), but the working conditions in many of these regions have always been difficult. Terraces are some of the most admired cultural landscapes in Mediterranean areas, but locals often use the expression "slavery land" to describe the harshness of the working conditions (Pereira et al. 2005).

Are Current Efforts to Maintain Traditional Landscapes Likely to Succeed?

Traditional agricultural practices were characterized by being labor intensive for relatively low agricultural yields (MacDonald et al. 2000; Gellrish et al. 2007). These characteristics played a key role in the demise of many of the traditional practices when labor costs rose due to economic growth, an effect that contributed to and was exacerbated by rural exodus. Large numbers of livestock kept vegetation succession on hold for centuries, but in the past few decades livestock numbers have declined in many of these regions (Cooper et al. 2006). In Europe, the number of livestock (cattle, goats and sheep) declined by 25 % between 1990 and 2010 (FAOSTAT 2010).

Still, recognizing the role of European farmers in maintaining these landscapes (Daugstad et al. 2006), several measures have been implemented to limit farmland depopulation. As part of the European Common Agriculture Policy, Less Favored Areas (LFAs-Regulation 1257/1999) were designated mainly to prevent rural abandonment and maintain cultural landscapes (Dax 2005; Stoate et al. 2009). LFAs went from representing a third of the European Utilized Agricultural Area (UAA) in 1975 to more than half in 2005 (Dax 2005; MacDonald et al. 2000). Though the LFA classification often happens to match High Nature Value farming systems and extensive agriculture, it poses no limit to intensification and overgrazing (Dax 2005).

In the Rural Development Plan for 2007–2013, the payments to farmers in LFAs totaled € 12.6 billion (DG Agriculture 2011). Though the sum of these subsidies is substantial at the European scale, at the individual level they might not be enough

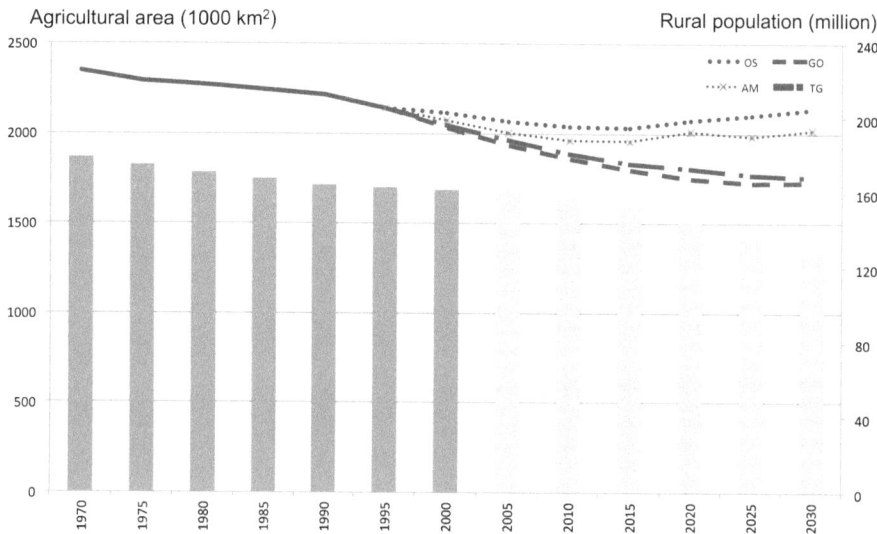

Fig. 1.3 Past and future trends of European agricultural area and rural population. Agricultural area (lines): land-use change predicted in the four scenarios of the Millennium Ecosystem Assessment (van Vuuren et al. 2006). The projections are based on the area of food crops, grass and fodder, and biofuels crops, between 1970 and 2030. OS order from strength, AM adapting mosaic, GO global orchestration, TG techno-garden. Rural population size (bars): historical values (*dark gray*) and future projections (*light gray*) (FAOSTAT 2010; past data for the Baltic countries from http://www.nationmaster.com)

to maintain young farmers or attract new residents (Cooper et al. 2006), especially in areas where the farm size is small. For example, when considering an average farm size of 23 ha in mountain areas (MacDonald et al. 2000) and an average LFA subsidy of € 100/ha (Dax 2005), the average payment is of € 2,300 per farm/year. This value can be higher if farmers also adhere to agri-environmental schemes, but overall LFA farmers still have lower incomes (Cooper et al. 2006): the Farm Net Value Added is 13,056 €/ Annual Work Unit in mountain LFAs, 14,174 €/AWU in other LFAs, and 18,923 €/AWU in non-LFAs (average for the EU25 countries between 2004 and 2005 in EC 2008b).

Hence the decrease in rural populations that started in the 1960s is projected to continue into the next few decades (Fig. 1.3). Future scenarios predict that the contribution of agriculture in regards to GDP and employment in Europe will continue decreasing (Eickhout et al. 2007; Nowicki et al. 2006) and the young generations will keep migrating to the cities, as long as their life quality and income prospects are higher there (EC 2008a; Keenleyside and Tucker 2010) resulting in the non-replacement of the aging population of European farmers.

Following the decrease in the rural population, agricultural area in Europe is also expected to keep contracting (Fig. 1.3), despite an expected increase in the global demand for agricultural goods, because enough food is obtained either directly by production on competitive land in Europe or elsewhere in the world (Keenleyside and Tucker 2010). Regionally labeled and organic products could help maintain

Table 1.1 Projections of future change in the agricultural area (arable land and pasture) from different studies

Region	Variation in the agricultural area	Initial agricultural area (Mha)	Period	Reference
EU15 + Norway and Switzerland[a]	−6%/−10% for cropland −1%/−10% for grassland	142.5	2000–2080	Rounsevell et al. 2006
EU15[b]	+5.5%/−15%	82.5	2000–2030	Eickhout et al. 2007
EU27	−5%/−15%	198	2000–2030	Verburg and Overmars 2009
Europe	−5%/−24%	235	1970–2050	MA 2005
Developed countries[c]	+8%/−20%	183	2000–2050	Balmford et al. 2005

[a] Initial agricultural area estimate obtained from FAOSTAT (2010)
[b] These values are only for arable land
[c] This study looked at the 23 most important food crops worldwide, corresponding to 44% of the cropland area in developed countries

certain forms of extensive agriculture but this market remains restricted (Strijker 2005). Projections also take into account an increasing demand in biocrops (Rounsevell et al. 2006; Schröter et al. 2005; Verburg and Overmars 2009), which can explain a moderate increase in the predicted agricultural area in some scenarios.

The dimension of the agricultural area abandoned or converted into production forest varies widely between scenarios (Table 1.1). If we use the intermediate scenarios in Verburg and Overmars (2009), between 10 and 29 million ha of land will be released from agriculture between 2000 and 2030. Areas particularly susceptible to the decline of agropastoral use include semi-natural grasslands and remote or mountainous areas with poor soil quality (Keenleyside and Tucker 2010; Pointereau et al. 2008; Stoate et al. 2009). Some of these areas are located in Northern Portugal, Northwestern France, the Alps, the Apennines and Central Europe (Fig. 1.4).

1.3 The Benefits of Rewilding

Defining Rewilding

Rewilding is the passive management of ecological succession with the goal of restoring natural ecosystem processes and reducing human control of landscapes (Gillson et al. 2011). Note that although passive management emphasizes no management or low levels of management (for example, Vera 2009), intervention may be required in the early restoration stages.

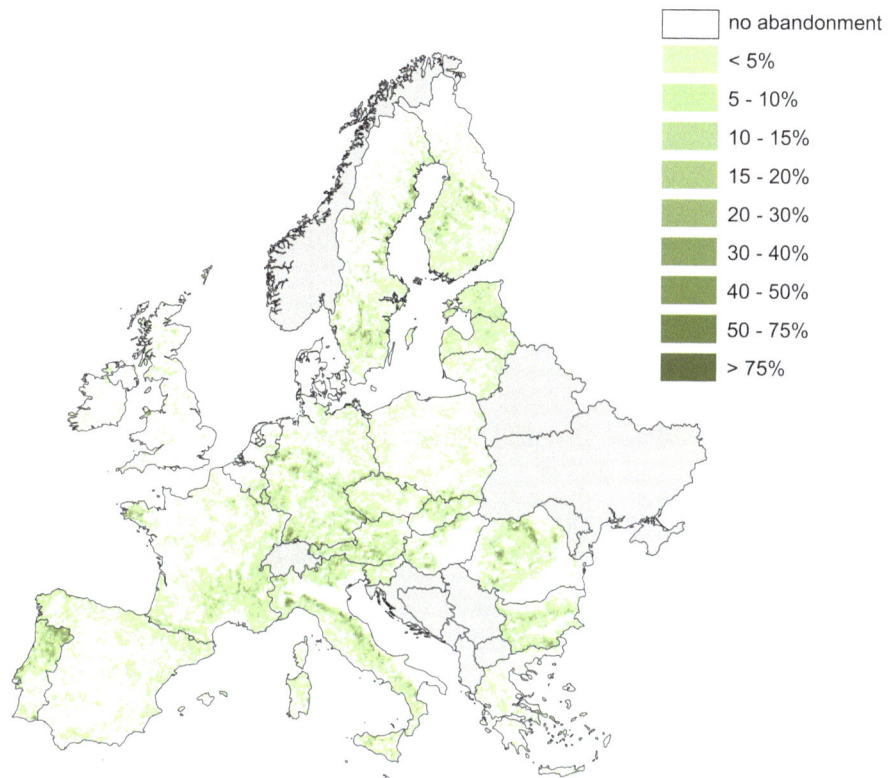

Fig. 1.4 Localization of the hotspots of abandonment and rewilding in Europe. Those hotspots are areas categorized as "agriculture" in 2000 that are projected to become rewilded or afforested in 2030 and that are common to all four scenarios of the CLUE model (Verburg and Overmars 2009). Hotspots are expressed as a percentage of each 100-km^2 grid cell. Agricultural areas correspond to "arable land (non-irrigated)", "pasture", "irrigated arable land" and "permanent crops". Rewilded and afforested areas correspond to "(semi)-natural vegetation", "forest", "recently abandoned arable land", and "recently abandoned pasture land". Countries in grey have no data

In contrast, much of the biodiversity conservation efforts in Europe emphasize active management, by maintaining low-level agricultural practices (Fig. 1.1). Active management also differs in goals, targeting the increase of the abundance of specific taxa or the maintenance of particular habitats, using approaches such as vegetation clearing and construction of artificial habitats, often working against successional processes.

Natural succession on abandoned farmland and pastures often leads to scrubland and sometimes at a later stage, to forest (Conti and Fagarazzi 2005). Passive forest regeneration restores almost as much forested areas globally as active tree plantation (Rey Benayas and Bullock 2012). Nonetheless, "wilderness" is not a synonym of "continuous forest" (Sutherland 2002). The European megafauna played a role in maintaining open landscapes, before being brought to global or local extinction by humans and replaced by domesticated grazers (Johnson 2009; Vera 2000; Bullock 2009, see Chap. 8).

This does not mean that rewilding should aim at rebuilding Pleistocene eco-systems, an approach which has been proposed elsewhere (Donlan et al. 2006), but that faces many difficulties (Caro 2007), including the lack of many of the original keystone species, a different climate, and ecosystems modified locally (for example, changes in soil caused by agriculture) and regionally by humans (for example, the global nitrogen cycle). Instead, the emphasis is on the development of self-sustaining ecosystems, protecting native biodiversity and natural ecological processes and providing a range of ecosystem services (Cramer et al. 2008). These novel ecosystems may be designed to be as similar as possible to some historical baseline in the recent or distant past, but they will often involve the introduction of new biotic elements (Hobbs et al. 2009).

Benefits of Rewilding for Biodiversity

Rewilding will cause biodiversity changes with some species declining in abundance, that is, loser species, and other species increasing in abundance, that is, winner spe-cies (Russo 2006; Sirami et al. 2008). We reviewed 23 studies identifying a positive response of species to decreasing human pressure or to restoration of their habitat following land abandonment (Supplementary Information[1], see also Chap. 4 to 8). In total, we identified 60 species of birds, 24 species of mammals, and 26 species of invertebrates that could benefit from farmland abandonment (Supplementary Table 1). We also identified 101 species negatively affected by land abandonment (Supplementary Table 2), but 13 of those species can be classified as both "winner" and "looser" depending on the study and the region. Much of the agrobiodiversity associated with High Nature Value Farmland will be in the "loosing" category. In contrast, many of the winner species have declined or became functionally extinct in traditional agricultural landscapes, such as large carnivores (see Chap. 4). These species will benefit from forest regeneration and the connection of fragmented natural habitats (Keenleyside and Tucker 2010; Russo 2006).

Revegetation promotes the increase of the organic matter content and the water holding capacity of soils (Arbelo et al. 2006). This can lead to higher biomasses and densities of earthworms (Russo 2006) and other invertebrate families (Supplementary Table 1.A).

Some forest birds benefit from forest regrowth after farmland abandonment (Pointereau et al. 2008), such as woodpeckers, treecreepers, and tits (Supplementary Table 1.B). Some birds of prey have benefited from increases in rodent popula-tions (Pointereau et al. 2008). Perhaps more surprisingly, populations of several bird species of the Eastern European steppe have increased after agricultural activity decline (Hölzel et al. 2002). Some, such as the Little Bustard (*Tetrax tetrax*), have benefited from the tall and dense grassland of the regrown steppes. This contrasts with the concerns that the decrease of open areas in Western Europe is contributing to the decline of steppe species. Therefore the biodiversity consequences of rewilding depend on the geographical context.

---

[1] http://link.springer.com/article/10.1007%2Fs10021-012-9558-7.

Likewise, rural abandonment makes the land suitable for a comeback of large mammals (Supplementary Table 1.C). Large grazers are benefiting from the lower hunting pressures that usually accompany abandonment (Breitenmoser 1998; Gortázar et al. 2000). European carnivore species have been increasing since the 1960s in abundance and distribution, as stable populations of Eastern Europe are naturally recolonizing abandoned landscapes of Scandinavia, the Mediterranean, and the Alps (Enserink and Vogel 2006; Boitani 2000; Stoate et al. 2009).

It is also important to consider the trophic interactions between species and the cascading effects driven by rewilding. For example, amphibians and otter (*Lutra lutra*) populations are known to benefit from the restoration of ditches by beavers (*Castor fiber*) in abandoned areas of Eastern Europe (Kull et al. 2004). The presence of lynx in some parts of Switzerland reduced the roe deer and chamois browsing impact by regulating both populations (Breitenmoser 1998).

Benefits of Rewilding for People: Ecosystem Services

Abandoned farmland is often perceived negatively as it is associated with the perception of unkept land and with the decrease on the economic usability of the land, particularly by the rural populations (Hochtl et al. 2005; Bauer et al. 2009). However there are many ecosystems services that are provided by this type of landscapes, particularly indirect and non-use services, which are often disregarded in the process of policy-making (TEEB 2010).

Rewilded areas can, at the regional scale, provide habitat for biodiversity with conservation results as high or higher than other land management options (Figs. 1.2, 1.5). This supporting service can lay the foundations for some cultural services (Fig. 1.5), because some of the species benefiting from abandonment are linked with recreation through hunting and tourism (Gortázar et al. 2000; Kaczennsly et al. 2004). For instance, in the Abbruze region of Italy, tourism has benefited from the advertisement of the presence of bears and wolves (Enserink and Vogel 2006). In addition to these direct and indirect use values, the large mammal species brought back by rewilding are amongst the species with highest existence values (Proença et al. 2008).

Forest regrowth promotes carbon sequestration (Kuemmerle et al. 2008). The carbon stock in European forests has grown from 5.3 to 7.7 PgC between 1950 and 1999 (Nabuurs et al. 2003). Nonetheless, active afforestation can potentially yield higher carbon sequestration rates than rewilding by using fast growing species (Fig. 1.5). Natural regeneration allows soil recovery and nutrient availability, though erosion can increase in the first years following abandonment (Pointereau et al. 2008; Rey Benayas et al. 2007). Forests regulate hydrological cycles, particularly in mountain areas (Körner et al. 2005) and water quality is expected to locally improve in abandoned fields (Stoate et al. 2009). Nonetheless, the transition from grassland to forest, a higher water-use system, can reduce the quantity of water (Brauman et al. 2007). Afforested areas managed for timber provisioning are disturbed both for plantation and management, thus providing qualitatively less water and soil related services than rewilded areas (Fig. 1.5).

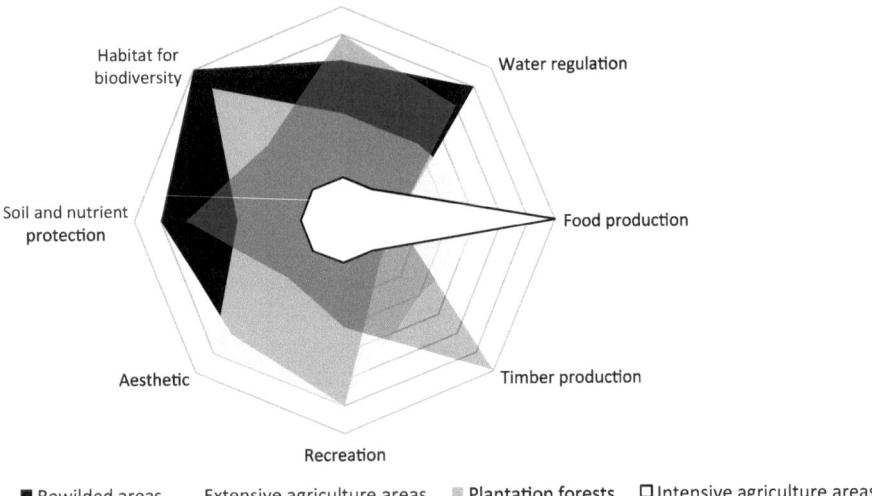

Fig. 1.5 Qualitative assessment of the ecosystem services provided by rewilding, afforestation, extensive agriculture and intensive agriculture in Europe. The relative values given to the provision of each service by the different land management strategies are discussed in the text

Intensive agriculture areas and planted forests are designed to focus on specific provisioning services. Extensive agriculture offers a tradeoff between food provisioning, cultural services, and habitat for biodiversity, whereas rewilding provides a wide range of supporting, regulating and cultural services (Fig. 1.5 and see Chap. 3).

The passive management associated with rewilding has much lower maintenance costs than other management options, and therefore significant returns of regulating and cultural services are obtained for limited levels of investment. Still, these services have characteristics of common goods (TEEB 2010), and therefore are rarely advantageous for the individual land-owner. Nonetheless, wilderness is linked to amenity-based growth and attracts urban individuals seeking different environments to both visit and work (Rasker and Hackman 1996): North American counties favoring wilderness showed faster growth in their employment and income level than counties in which the economy is mainly based on resource extraction.

1.4 The Challenges of Rewilding

Rewilding as a landscape management option does involve several challenges. Our understanding of those challenges and how they can be overcome depends on the relationship between humans, the landscape and the biodiversity that it sustains.

Conflicts with Wildlife

Conflicts occur when wildlife overlaps with human activities such as hunting and farming (Gortázar et al. 2000; Linnell et al. 2000; Schley and Roper 2003). Those conflicts are age-old in Europe and negative perceptions were transmitted through generations via folklore and tales (Wilson 2004; Boitani 2000). Hunting wild species, and particularly carnivores, was socially enforced (Enserink and Vogel 2006), which led in many cases to their local extinction by the nineteenth century.

Though many European countries have implemented regulations to protect large carnivores, such legislation is not understood and accepted by all (Breitenmoser 1998). In particular, they accentuate a cleavage in opinions amongst countries and between rural and urban populations (Bauer et al. 2009; Wilson 2004) the latter being usually more favorable to a wildlife comeback.

The conflicts with carnivores are largely explained by the fact that they prey on domestic animals due to the scarcity of wild prey (Russo 2006) but also by the loss of traditional livestock-guarding knowledge in several countries (Fourli 1999; Kaczensky et al. 2004). Nonetheless, the level of depredation of livestock by carnivores is generally low, often less than 10 % of their diet (Wilson 2004). Still, the impact at the level of the livestock owner can be high (Wilson 2004). To compensate for these impacts, several countries pay for damages caused by wildlife. For bear and wolf damages, an average of € 2 million/year were compensated in Europe between 1992 and 1998 in France, Greece, Italy, Austria, Spain and Portugal (Fourli 1999) while € 2.15 million were spent in preventive measures.

Large grazers such as deer and wild boars can also cause significant damage to crops, pastures and forest plantations (Goulding and Roper 2002; Kamler et al. 2010). As for the carnivores, a combination of preventive measures such as electric fencing (Honda et al. 2009) with compensation payments can contribute to decrease the levels of conflict.

Fear of attacks on people also play a factor in this conflict, but this often can be improved with better information to the public as there is a correlation between the fear of an animal and a lack of knowledge of its behavior (Decker et al. 2010; Kaczensky et al. 2004).

Limits to Ecological Resilience

In many regions of Europe, the transition from abandoned to semi-natural land takes less than 15 years, followed by another 15–30 years before reforestation (Cramer et al. 2008; Verburg and Overmars 2009). Passive regeneration can therefore be a slow process, particularly in a dry environment such as the Mediterranean (Rey Benayas et al. 2008), or when the soils have been modified by past agriculture, that is, the "cultivation legacy" (Cramer et al. 2008), or the "grazing history" (Chauchard et al. 2007). The revegetation also depends on the availability and quality of the native seed bank (Rey Benayas et al. 2008).

If the abandoned land is too degraded assisted regeneration may be needed (Cramer et al. 2008). Active restoration would involve large-scale native trees plantation and tree growth management (Rey Benayas et al. 2008). An intermediate level of intervention involves the creation and management of forest regeneration sources or "woodland islets" (Rey Benayas and Bullock 2012; see Chap. 7). Another problem often requiring intervention is the vulnerability of intermediate stages of natural succession to natural perturbations, such as invasive species (Kull et al. 2004; Stoate et al. 2009) and fire (Pausas et al. 2008). Fire is a particularly acute problem as it has impacts not only on biodiversity but also on human health (Proença and Pereira 2010b). If fire regime is not appropriately managed, frequent fires will favor fire-prone scrubland and halt succession towards forest, in a self-reinforcing feedback loop (Proença and Pereira 2010a).

One of the strategies to manage fire regimes is to maintain open spaces in the landscape (see Chap. 8), minimizing also the impacts of revegetation on species that prefer open areas (Fig. 1.2). This strategy can be implemented by increasing the populations of large herbivores (Hodder and Bullock 2009; Sutherland 2002), including reintroduction of extinct species (Svenning 2002). In the case of species regionally extinct, it is possible to use individuals from other populations. For instance, seven European bison were recently reintroduced in northern Spain, 1,000 years after their extinction (Burton 2011). A more complex situation occurs with species that are globally extinct, such as wild relatives of some domesticated species. A possible solution is to release into the wild individuals of breeds that are most likely to be successful in replacing the ecological role of their wild ancestors. For instance, Iceland ponies have been released in the former arable fields of the Dutch-Belgian border (Kuiters and Slim 2003): their grazing favored a dense grass sward and after 27 years open grassland still represented 98 % of the area.

Natural colonization of abandoned land by carnivores can also be limited by the availability of prey, as is the case for the Iberian lynx (*Lynx pardinus*) currently negatively affected by the scarcity of rabbits, decimated by diseases (Delibes-Mateos et al. 2008), or as can be expected for some populations of wolves and bears currently preying on livestock (Russo 2006).

Rewilding may be a future option in areas that are undergoing agricultural development or intensification today. There is currently a debate between land sharing and land sparing approaches to reconcile food production with biodiversity (Phalan et al. 2011). In land sharing, biodiversity conservation and food production goals are met on the same land, with biodiversity friendly agricultural practices and extensive agriculture, whereas in land sparing, land is divided between areas of intensification and of exclusion of agriculture. In practice, it is difficult to determine which is the best option because species respond differently to the alteration of their habitat (Phalan et al. 2011). To maintain future options for rewilding, both land sparing and land sharing are needed. On the one hand, land sharing is essential to limit land degradation and to maintain the appropriate seed bank for future passive revegetation (see Chap. 7). On the other hand, land sparing would allow for the conservation of populations of species that are currently in conflict with human activities, making "cohabitation" very difficult.

1.5 Final Remarks

Most landscapes are evaluated and protected according to emotional and aesthetic values that societies attribute to them (Antrop 2005; Gobster et al. 2007) and conservation programs are determined by people's perceptions of what should be preserved (Gillson et al. 2011) and depend on shifting baselines of what nature should be like (Vera 2009). Thus, the values that Europeans give to farmland and wilderness landscapes are based on tradition and history but also on socio-economic backgrounds (Van den Berg and Koole 2006). Yet, considering that landscapes result from the dynamic interaction of natural and cultural drivers (Antrop 2005), they cannot be perceived as anchored in time and we should anticipate occasional changes that will force us to reevaluate their definition.

Rewilding appears to be a viable management option for some of these transitions with important benefits for biodiversity and ecosystem services. At the local scale, some species will decline and other increase, eventually leading to local species diversity decreases in some taxa (Fig. 1.2). We lack research studies looking at the regional scale dynamics, but we hypothesize that no significant loss in species diversity is expected as long as mosaics of open spaces and forest are maintained, and that some dimensions of biodiversity may even improve, such as the average size of populations of wild species. At the global scale, many species have already gone extinct and it will be impossible to get them back, but the release into the wild of breeds of some domesticated species may allow recovery of some historical losses (Fig. 1.2). In terms of ecosystem services, rewilding allows for a wide range of regulating and cultural services (Fig. 1.5).

The extent and outcome of rewilding will be heterogeneous across Europe (Fig. 1.4) as different regions will have different departing points of post-farmland abandonment and varying limitations to natural forest regrowth. For example, on some abandoned areas of Southern Europe, the availability of forest tree seed banks can be a limiting factor due to little natural forest left and the frequent fire regime may delay ecological succession. In contrast, the relative scarcity of open areas in much of Northern Europe may render the intensification or reestablishment of natural perturbations, such as grazing by large wild herbivores and fire (for example, prescribed burns), priority goals for management. Rewilding can also be considered on available land that does not necessarily result from farmland abandonment, such as national forests previously managed for timber production, decommissioned military areas, salt ponds and other wetlands, thus increasing the level of heterogeneity of European wild landscapes.

From a conservation standpoint, the option between rewilding and active management will depend on the goals and the local context. Active management is likely to be preferred when the goal is to restore specific species or maintain early successional habitats and other habitats associated with human activities. Passive management emphasizes dynamic ecological processes over static patterns of species or habitat occurrence and can be more sustainable in the long term or at large spatial scales.

Despite many benefits, rewilding has been disregarded as a management option until recently. Initiatives such as Rewilding Europe (http://www.rewildingeurope. com, and see Chap. 9) and the PAN Parks Network (http://www.panparks.org) are now bringing rewilding to the forefront of the discussion of European conservation policies. Rewilding poses many challenges, but those are inherent to the implementation of any restoration plan. In a world wounded by biodiversity loss, farmland abandonment is an opportunity to improve biodiversity in Europe, to study the regeneration of vegetation, and even to test ecological theories (Hobbs and Cramer 2007). In the end, the question is not whether we prefer a domesticated or a wild European landscape but rather which management options (Fig. 1.1) at each place will be more achievable and sustainable.

Acknowledgments We thank P. Verburg for sharing data from the CLUE model and commenting on the manuscript. We also thank V. Proença, R. Beilin, J. Bullock and S. Ceauşu for comments. This research was funded by the Fundação para a Ciência e a Tecnologia (FCT)-ABAFOBIO (PTDC/AMB/73901/2006) and by FORMAS-Project LUPA. L.N. was supported by a grant from FCT (SFRH/BD/62547/2009).

References

Antrop, M. (2005). Why landscapes of the past are important for the future. *Landscape and Urban Planning, 70,* 21–34.

Arbelo, C. D., Rodríguez-Rodríguez, A., Guerra, J. A., Mora, J. L., Notario, J. S., & Fuentes, F. (2006). Soil degradation processes and plant colonization in abandoned terraced fields overlying pumice tuffs. *Land Degradation and Development, 17,* 571–588.

Balmford, A., Green, R. E., & Scharlemann, J. P. W. (2005). Sparing land for nature: Exploring the potential impact of changes in agricultural yield on the area needed for crop production. *Global Change Biology, 11,* 1594–1605.

Batáry, P., Holzschuh, A., Orci, K. M., Samu, F., & Tscharntke, T. (2012). Responses of plant, insect and spider biodiversity to local and landscape scale management intensity in cereal crops and grasslands. *Agriculture, Ecosystems & Environment, 146*(1), 130–136.

Bauer, N., Wallner, A., & Hunziker, M. (2009). The change of European landscapes: Human-nature relationships, public attitudes towards rewilding, and the implications for landscape management in Switzerland. *Journal of Environmental Management, 90,* 2910–2920.

Bell, S., Montarzino, A., Aspinall, P., Peněze, Z., & Nikodemus, O. (2009). Rural society, social inclusion and landscape change in Central and Eastern Europe: A case study of Latvia. *European Society for Rural Sociology, 49,* 295–326.

Blondel, J. (2006). The "design" of mediterranean landscapes: A millennial story of humans and ecological systems during the historic period. *Human Ecology, 34,* 713–729.

Boitani, L. (2000). *Action plan for the conservation of the wolves (Canis lupus) in Europe. nature and environment, no. 113* (p. 84). Strasbourg: Council of Europe Publishing.

Brauman, K. A., Daily, G. C., Duarte, T. K., & Mooney, H. A. (2007). The nature and value of ecosystem services: An overview highlighting hydrologic services. *Annual Review of Environment and Resources, 32,* 67–98.

Breitenmoser, U. (1998). Large predators in the Alps: The fall and rise of man's competitors. *Biological Conservation, 83*, 279–289.

Bugalho, M. N., Caldeira, M. C., Pereira, J. S., Aronson, J., & Pausas, J. G. (2011). Mediterranean Cork Oak Savannas require human use to sustain biodiversity and ecosystem services. *Frontiers in Ecology and Environment, 9*(5), 278–286.

Bullock, D. J. (2009). What larger mammals did Britain have and what did they do? *British Wildlife, 20*(5), 16–20.

Burton, A. (2011). Where the wisents roam. *Frontiers in Ecology and Environment, 9*, 140.

Caro, T. (2007). The Pleistocene re-wilding gambit. *Trends in Ecology & Evolution (Personal edition), 22*, 281–283.

Chauchard, S., Carcaillet, C., & Guibal, F. (2007). Patterns of land-use abandonment control tree-recruitment and forest dynamics in Mediterranean mountains. *Ecosystems, 10*, 936–948.

Conti, G., & Fagarazzi, L. (2005). Forest expansion in mountain ecosystems: "environmentalist's dream" or societal nightmare? *Planum, 11*, 1–20.

Cooper, T., Baldock, D., Rayment, M., Kuhmonen, T., Terluin, I., Swales, V., Poux, X., Zakeossian, D., & Farmer, M. (2006). *An evaluation of the less favoured area measure in the 25 member states of the European Union* (p. 262). London: Institute for European Environmental Policy.

Cramer, V. A., Hobbs, R. J., & Standish, R. J. (2008). What's new about old fields? Land abandonment and ecosystem assembly. *Trends in Ecology & Evolution (Personal edition), 23*, 104–112.

Daugstad, K., Ronningen, K., & Skar, B. (2006). Agriculture as an upholder of cultural heritage? Conceptualizations and value judgements—a Norwegian perspective in international context. *Journal of Rural Studies, 22*, 67–81.

Dax, T. (2005). The redefinition of Europe's less favoured areas. In Rural development in Europe—3rd annual conference—Funding European Rural Development in 2007–2013. MPRA paper no. 711.

Decker, S. E., Bath, A. J., Simms, A., Lindner, U., & Reisinger, E. (2010). The return of the king or bringing snails to the garden? The human dimensions of a proposed restoration of European Bison (*Bison bonasus*) in Germany. *Restoration Ecology, 18*, 41–51.

Delibes-Mateos, M., Delibes, M., Ferreras, P., & Villafuerte, R. (2008). Key role of European rabbits in the conservation of the Western Mediterranean basin hotspot. *Conservation Biology: The Journal of the Society for Conservation Biology, 22*(5), 1106–1117.

Devy-Vareta, N., & Alves, A. A. M. (2007). Os avanços e os recuos da floresta em Portugal-da Idade Média ao Liberalismo. In J. S. Silva (Ed.), *Floresta e sociedade, uma historia em comum* (pp 55–75). Lisboa: Público SA e Fundação Luso-Americana.

Agriculture, D. G. (2011). Rural development in the European Union. Statistical and economic information report (p. 257).

Donlan, C. J., Berger, J., Bock, C. E., Bock, J. H., Burney, D. A., Estes, J. A., Foreman, D., Martin, P. S., Roemer, G. W., Smith, F. A., Soulé, M. E., & Greene, H. W. (2006). Pleistocene rewilding: An optimistic agenda for twenty-first century conservation. *The American Naturalist, 168*(5), 660–681.

EEA (2004). *High nature value farmland: Characteristics, trends and policy challenges* (p. 31). Copenhagen: European Environmental Agency.

EC—European Commission. (2008a). *Poverty and social exclusion in rural areas* (p. 187). Brussels: DG Employment Social Affairs and Equal Opportunities.

EC—European Commission. (2008b). *Overview of the less favoured areas farms in the EU-25 (2004–2005)* (p. 99). Brussels: DG Agriculture and Rural Development.

EC—European Commission. (2009). *New insights into mountain farming in the European Union* (p. 35). Brussels: DG Agriculture and Rural Development.

Eickhout, B., Van Meijl H., Tabeau, A., & Van Rheenen, T. (2007). Economic and ecological consequences of four European land use scenarios. *Land Use Policy, 24*, 562–575.

Enserink, M., & Vogel, G. (2006). The carnivore comeback. *Science, 314*, 746–749.

Farrell, E. P., Führer, E., Ryan, D., Andersson, F., Hüttl, R., & Piussi, P. (2000). European forest ecosystems: Building the future on the legacy of the past. *Forest Ecology and Management, 132*, 5–20.

FAO. (2011). *State of the world's forests* (p. 179). Rome: FAO.

FAOSTAT. (2010). http://faostat.fao.org. Accessed: 1. March 2011.

Figueiredo, J., & Pereira, H. M. (2011). Regime shifts in a socio-ecological model of farmland abandonment. *Landscape Ecology, 26*(5), 737–749.

Fourli, M. (1999). Compensation for damage caused by bears and wolves in the European Union. LIFE-Nature projects, European Commission-DG XI-Environment, Nuclear Safety and Civil Protection. 72p. Brussels.

Gellrich, M., Baur, P., Koch, B., & Zimmermann, N. E. (2007). Agricultural land abandonment and natural forest re-growth in the Swiss mountains: A spatially explicit economic analysis. *Agriculture, Ecosystems & Environment, 118*, 93–108.

Gillson, L., Ladle, R. J., & Araújo, M. B. (2011). Baselines, patterns and process. In R. J. Ladle, R. J. Whittaker (Eds.), *Conservation biogeography* (pp. 31–44). Oxford: Wiley-Blackwell.

Gobster, P. H., Nassauer, J. I., Daniel, T. C., & Fry, G. (2007). The shared landscape: What does aesthetics have to do with ecology? *Landscape Ecology, 22*, 959–972.

Gortázar, C., Herrero, J., Villafuerte, R., & Marco, J. (2000). Historical examination of the status of large mammals in Aragon, Spain. *Mammalia, 64*, 411–422.

Goulding, M. J., & Roper, T. J. (2002). Press responses to the presence of free-living wild boar (Sus scrofa) in southern England. *Mammal Reviews, 32*, 272–282.

Halada, L., Evans, D., Romão, C., & Petersen, J.-E. (2011). Which habitats of European importance depend on agricultural practices? *Biodiversity and Conservation, 20*(11), 2365–2378.

Hobbs, R. J., & Cramer, V. A. (2007). Why old fields? Socioeconomic and ecological causes and consequences of land abandonment. In V. A. Cramer, R. J. Hobbs (Eds.), *Old fields: Dynamic and restoration of abandoned farmland* (pp. 1–14). Washington: Island Press.

Hobbs, R. J., Higgs, E., & Harris, J. A. (2009). Novel ecosystems: Implications for conservation and restoration. *Trends in ecology & evolution (Personal edition), 24*, 599–605.

Höchtl, F., Lehringer, S., & Konold, W. (2005). "Wilderness": What it means when it becomes a reality—a case study from the southwestern Alps. *Landscape and Urban Planning, 70*, 85–95.

Hodder, K. H., & Bullock, J. M. (2009). Really wild? Naturalistic grazing in modern landscapes. *British Wildlife, 20*, 37–43.

Hodder, K. H., Buckland, P. C., Kirby, K. K., & Bullock, J. M. (2009). Can the pre-neolithic provide suitable models for re-wilding the landscape in Britain? *British Wildlife, 20*(5), 4–15.

Hölzel, N., Haub, C., Ingelfinger, M. P., Otte, A., & Pilipenko, V. N. (2002). The return of the steppe—large-scale restoration of degraded land in southern Russia during the post-Soviet era. *Journal for Nature Conservation, 10*, 75–85.

Honda, T., Miyagawa, Y., Ueda, H., & Inoue, M. (2009). Effectiveness of newly-designed electric fences in reducing crop damage by medium and large mammals. *Mammal Study, 34*, 13–17.

Johnson, C. N. (2009). Ecological consequences of late quaternary extinctions of megafauna. *Proceedings of the Royal Society B, 276*, 2509–2519.

Kaczensky, P., Blazic, M., Gossow, H. (2004). Public attitudes towards brown bears (*Ursus arctos*) in Slovenia. *Biological Conservation, 118*, 661–674.

Kamler, J., Homolka, M., Barancěková, M., & Krojerová-Prokesõvá, J. (2010). Reduction of herbivore density as a tool for reduction of herbivore browsing on palatable tree species. *European Journal of Forest Research, 129*, 155–162.

Kaplan, J. O., Krumhardt, K. M., & Zimmermann, N. (2009). The prehistoric and preindustrial deforestation of Europe. *Quaternary Science Reviews, 28*, 3016–3034.

Keenleyside, C., & Tucker, G. 2010. *Farmland Abandonment in the EU: An assessment of trends and prospects* (p. 97) London: WWF and IEEP.

Körner, C., Spehn, E., & Baron, J. (2005). *Mountain systems. Millenium ecosystem assessment. Ecosystems and human well-being: Current state and trends* (pp. 681–716). Washington: Island Press.

Kuemmerle, T., Hostert, P., Radeloff, V. C., Linden, S., Perzanowski, K., & Kruhlov, I. (2008). Cross-border comparison of post-socialist farmland abandonment in the Carpathians. *Ecosystems, 11*, 614–628.

Kuiters, A. T., & Slim, P. A. (2003). Tree colonisation of abandoned arable land after 27 years of horse-grazing: The role of bramble as a facilitator of oak wood regeneration. *Forest Ecology Management, 181,* 239–251.

Kull, T., Pencheva, V., Petrovic, F., Elias, P., Henle, K., Balciauskas, L., Kopacz, M., Zajickova, Z., & Stoianovici, V. (2004). Agricultural landscapes. In J Young, L Halada, T Kull, A Kuzniar, U Tartes, Y Uzunov, & A Watt (Eds.), *Conflicts between human activities and the conservation of biodiversity in agricultural landscapes, grasslands, forests, wetlands and uplands in the acceding and candidate countries* (pp. 10–20) Wallingford: Centre for Ecology and Hydrology.

Laiolo, P., Dondero, F., Ciliento, E., & Rolando, A. (2004). Consequences of pastoral abandonment for the structure and diversity of the alpine avifauna. *Journal of Applied Ecology, 41,* 294–304.

Lindborg, R., Bengtsson, J., Berg, A., Cousins, S. A. O., Eriksson, O., Gustafsson, T., Hasund, K. P., Lenoir, L., Pihlgren, A., Sjödin, E., & Stenseke, M. (2008). A landscape perspective on conservation of semi-natural grasslands. *Agricultural and Ecosystem Environment, 125*(1), 213–222.

Linnell, J. D. C., Swenson, J. E., & Andersen, R. (2000). Conservation of biodiversity in Scandinavian boreal forests: Large carnivores as flagships, umbrellas, indicators, or keystones? *Biodiversity and Conservation, 9,* 857–868.

MA—Millennium Ecosystem Assessment. (2005). *Ecosystems and human well-being: Scenarios.* (p. 560). Washington: Island Press.

MacDonald, D., Crabtree, J. R., Wiesinger, G., Dax, T., Stamou, N., Fleury, P., Gutierrez Lazpita, J., & Gibon, A. (2000). Agricultural abandonment in mountain areas of Europe: Environmental consequences and policy response. *Journal of Environmental Management, 59,* 47–69.

McNeely, J. A. (1994). Lessons from the past: Forests and biodiversity. *Biodiversity and Conservation, 3,* 3–20.

Meijaard, E., & Sheil, D. (2011). A modest proposal for wealthy countries to reforest their land for the common good. *Biotropica, 43*(5), 524–528.

Moreira, F., & Russo, D. (2007). Modelling the impact of agricultural abandonment and wildfires on vertebrate diversity in Mediterranean Europe. *Landscape Ecology, 22,* 1461–1476.

Nabuurs, G. J., Schelhaas, M. J., Mohren, G. M. J., & Field, C. B. (2003). Temporal evolution of the European forest sector carbon sink from 1950 to 1999. *Global Change Biology, 9,* 152–160.

Nowicki, P., Weeger, C., Van Meijl H., Banse, M., Helming, J., Terluin, I., Verhoog, D., Overmars, K. P., & Westhoek, H. 2006. *SCENAR 2020: Scenario study on agriculture and the rural world* (p. 236). Brussels: European Commission-DG Agriculture and Rural Development.

Pausas, J. G., Llovet, J., Rodrigo, A., & Vallejo, R. (2008). Are wildfires a disaster in the Mediterranean basin—a review. *International Journal of Wildland Fire, 17,* 713–723.

Pereira, E., Queiroz, C., Pereira, H. M., & Vicente, L. (2005). Ecosystem services and human well-being: A participatory study in a mountain community in Portugal. *Ecology and Society, 10*(2), 14.

Pereira, H. M., Leadley, P. W., Proença, V., Alkemade, R., Scharlemann, J. P. W., Fernandez-Manjarrés, J. F., Araújo, M. B., Balvanera, P., Biggs, R., Cheung, W. W. L., Chini, L., Cooper, H. D., Gilman, E. L., Guénette, S., Hurtt, G. C., Huntington, H. P., Mace, G. M., Oberdorff, T., Revenga, C., Rodrigues, P., Scholes, R. J., Sumaila, U. R., & Walpole, M. (2010). Scenarios for global biodiversity in the 21st century. *Science, 330,* 1496–1501.

Pereira, H. M., Navarro, L. M., & Martins, I. S. 2012. Global biodiversity change: The good, the bad and the unknown. *Annual Review of Environment and resources, 37,* 25–50.

Phalan, B., Onial, M., Balmford, A., & Green, R. E. (2011). Reconciling food production and biodiversity conservation: Land sharing and land sparing compared. *Science, 333*(6047), 1289–1291.

Pinto-Correia, T., & Mascarenhas, J. (1999). Contribution to the extensification/intensification debate: New trends in the Portuguese Montado. *Landscape and Urban Planning, 46,* 125–131.

Pointereau, P., Coulon, F., Lambotte, M., Stuczynski, T., Sanchez Ortega, V., & Del Rio, A. (2008). *Analysis of farmland abandonment and the extent and location of agricultural areas that are actually abandoned or are in risk to be abandoned* (p. 204). Ispra: European Commission-JRC-Institute for Environment and Sustainability.

Proença, V., & Pereira, H. M. (2010a). Appendix 2: Mediterranean forest (pp. 60–67). In Leadley, P., Pereira, H.M., Alkemade, R., Fernandez-Manjarrés, J.F., Proença, V., Scharlemann, J.P.W., Walpole, M.J. (Eds.) *Biodiversity Scenarios: Projections of 21st century change in biodiversity and associated ecosystem services*. Secretariat of the Convention on Biological Diversity, Montreal. Technical Series no. 50, 132 pages.

Proença, V., & Pereira, H. M. (2010b). Ecosystem changes, biodiversity loss and human well-being. In J. O. Nriagu (Ed.), *Encyclopedia of environmental health* (pp. 215–224). Burlington: Elsevier.

Proença, V., Pereira, H. M., & Vicente, L. (2008). Organismal complexity is an indicator of species existence value. *Frontiers in Ecology Environment, 6*, 298–299.

Rasker, R., & Hackman, A. (1996). Economic development and the conservation of large carnivores. *Conservation Biology: The Journal of the Society for Conservation Biology, 10*, 991–1002.

Rey Benayas, J., & Bullock, J. (2012). Restoration of biodiversity and ecosystem services on agricultural land. *Ecosystems, 15*, 883–899.

Rey Benayas, J. M., Martins, A., Nicolau, J. M., & Schulz, J. J. (2007). Abandonment of agricultural land: An overview of drivers and consequences. *CAB Reviews, 2*, 1–14.

Rey Benayas, J. M., Bullock, J. M., & Newton, A. C. (2008). Creating woodland islets to reconcile ecological restoration, conservation, and agricultural land use. *Frontiers in Ecology Environment, 6*, 329–336.

Rounsevell, M. D. A., Reginster, I., Araújo, M. B., Carter, T. R., Dendoncker, N., Ewert, F., House, J. I., Kankaanpaa, S., Leemans, R., Metzger, M. J., Schmit, C., Smith, P., & Tuck, G. (2006). A coherent set of future land use change scenarios for Europe. *Agriculture, Ecosystems and Environment, 114*, 57–68.

Ruben, R., & Pender, J. (2004). Rural diversity and heterogeneity in less-favoured areas: The quest for policy targeting. *Food Policy, 29*, 303–320.

Russo, D. (2006). Effects of land abandonment on animal species in Europe: Conservation and management implications. Integrated assessment of vulnerable ecosystems under global change in the EU (p. 52). Project report.

Schley, L., & Roper, T. J. (2003). Diet of wild boar Sus scrofa in Western Europe, with particular reference to consumption of agricultural crops. *Mammal Review, 33*, 43–56.

Schröter, D., Cramer, W., Leemans, R., Prentice, I. C., Araújo, M. B., Arnell, N. W., Bondeau, A., Bugmann, H., Carter, T. R., & Gracia, C. A. (2005). Ecosystem service supply and vulnerability to global change in Europe. *Science, 310*, 1333–1337.

Sirami, C., Brotons, L., Burfield, I., Fonderflick, J., & Martin, J. L. (2008). Is land abandonment having an impact on biodiversity? A metaanalytical approach to bird distribution changes in the northwestern Mediterranean. *Biological Conservation, 141*, 450–459.

Stoate, C., Báldi, A., Beja, P., Boatman, N. D., Herzon, I., Van Doorn A., De Snoo, G. R., Rakosy, L., & Ramwell, C. (2009). Ecological impacts of early 21st century agricultural change in Europe—a review. *Journal of Environmental Management, 91*, 22–46.

Strijker, D. (2005). Marginal lands in Europe–causes of decline. *Basic and Applied Ecology, 6*, 99–106.

Sutherland, W. J. (2002). Openness in management. *Nature, 418*, 834–835.

Svenning, J. C. (2002). A review of natural vegetation openness in North-western Europe. *Biological Conservation, 104*(2), 133–148.

TEEB. (2010). The economics of ecosystems and biodiversity: Mainstreaming the economics of nature: A synthesis of the approach, conclusions and recommendations of TEEB, 39p.

Tscharntke, T., Klein, A. M., Kruess, A., Steffan-Dewenter, I., & Thies, C. (2005). Landscape perspectives on agricultural intensification and biodiversity—ecosystem service management. *Ecology Letters, 8*(8), 857–874.

Van den Berg, A. E., & Koole, S. L. (2006). New wilderness in the Netherlands: An investigation of visual preferences for nature development landscapes. *Landscape and Urban Planning, 78*, 362–372.

Van Vuuren, D. P., Sala, O. E., & Pereira, H. M. (2006). The future of vascular plant diversity under four global scenarios. *Ecology and Society, 11*(2), 25.

Vera, F. W. M. (2000). *Grazing ecology and forest history* (p. 527). New York: CABI.

Vera, F. W. M. (2009). Large-scale nature development—the Oostvaardersplassen. *British Wildlife, 20*(5), 28–36.

Verburg, P. H., & Overmars, K. P. (2009). Combining top-down and bottom-up dynamics in land use modeling: Exploring the future of abandoned farmlands in Europe with the Dyna-CLUE model. *Landscape Ecology, 24,* 1167–1181.

Williams, M. (2000). Dark ages and dark areas: Global deforestation in the deep past. *Journal of Historical Geography, 26,* 28–46.

Wilkinson, D. M. (1999). The disturbing history of intermediate disturbance. *Oikos, 84,* 145–147.

Wilson, C. J. (2004). Could we live with reintroduced large carnivores in the UK? *Mammal Review, 34,* 211–232.

Young, J., Watt, A., Nowicki, P., Alard, D., Clitherow, J., Henle, K., Johnson, R., Laczko, E., McCracken, D., Matouch, S., Niemela, J., & Richards, C. (2005). Towards sustainable land use: Identifying and managing the conflicts between human activities and biodiversity conservation in Europe. *Biodiversity Conservation, 14,* 1641–1661.

Chapter 2
European Wilderness in a Time of Farmland Abandonment

Silvia Ceauşu, Steve Carver, Peter H. Verburg, Helga U. Kuechly, Franz Hölker, Lluis Brotons and Henrique M. Pereira

Abstract Wilderness is a multidimensional concept that has evolved from an aesthetic idea to a science-based conservation approach. We analyze here several subjective and ecological dimensions of wilderness in Europe: human access from roads and settlements, impact of artificial night light, deviation from potential

S. Ceauşu (✉) · H. M. Pereira
German Centre for Integrative Biodiversity Research (iDiv) Halle-Jena-Leipzig
Deutscher Platz 5e, 04103 Leipzig, Germany
e-mail: silvia.ceausu@mespom.eu

Institute of Biology, Martin Luther University Halle-Wittenberg
Am Kirchtor 1, 06108 Halle (Saale), Germany

Centro de Biologia Ambiental, Faculdade de Ciências da Universidade de Lisboa
Campo Grande, 1749-016 Lisboa, Portugal

H. M. Pereira
e-mail: hpereira@idiv.de

S. Carver
Wildland Research Institute, School of Geography, University of Leeds, LS2 9JT, Leeds, UK
e-mail: S.J.Carver@leeds.ac.uk

P. H. Verburg
Institute for Environmental Studies (IVM), VU University Amsterdam
De Boelelaan 1087, 1081 HV, Amsterdam, The Netherlands
e-mail: peter.verburg@vu.nl

H. U. Kuechly · F. Hölker
Leibniz Institute of Freshwater Ecology and Inland Fisheries
Müggelseedamm 310, 12587 Berlin, Germany
e-mail: kuechly@posteo.de

F. Hölker
e-mail: hoelker@igb-berlin.de

L. Brotons
European Bird Census Council (EBCC), Centre de Recerca Ecològica i Aplicacions
Forestals (CREAF), Centre Tecnològic Forestal de Catalunya (CEMFOR—CTFC).
Ctra. antiga St. Llorenç km 2, 25280 Solsona, Spain
e-mail: lluis.brotons@ctfc.cat

H. M. Pereira, L. M. Navarro (eds.), *Rewilding European Landscapes,*
DOI 10.1007/978-3-319-12039-3_2, © The Author(s) 2015

natural vegetation and proportion of harvested primary productivity. As expected, high wilderness in Europe is concentrated mainly in low primary productivity areas at high latitudes and in mountainous regions. The use of various wilderness metrics also reveals additional aspects, allowing the identification of regional differences in the types of human impact and a better understanding of future modifications of wilderness values in the context of land-use change. This is because farmland abandonment in the next decades is projected to occur especially at intermediate wilderness values in marginal agricultural landscapes, and thus can release additional areas for wild ecosystems. Although the subjective wilderness experience will likely improve at a slower pace due to the long-term persistence of infrastructures, the ecological effects of higher resource availability and landscape connectivity will have direct positive impacts on wildlife. Positive correlation between megafauna species richness and wilderness indicate that they spatially coincide and for abandoned areas close to high wilderness areas, these species can provide source populations for the recovery of the European biota. Challenges remain in bringing together different views on rewilding and in deciding the best management approach for expanding wilderness on the continent. However the prospects are positive for the growth of self-regulating ecosystems, natural ecological processes and the wilderness experience in Europe.

Keywords Wilderness · Human footprint · Artificial light · Potential natural vegetation · Harvested primary productivity · Megafauna · Farmland abandonment

2.1   The History and Value of Wilderness

Wilderness is a comprehensive measure of conservation value capturing both the subjective human experience, and the ecological dimension of minimally impacted ecosystems (Cole and Landres 1996; Hochtl et al. 2005). But the concept of wilderness has gone through dramatic historical changes in terms of both the context and connotation in which the term was used. During the centuries of exploration and colonization of new territories, wilderness was perceived negatively as a land that is unfavourable for human habitation and should be altered and tamed (Nash 2001).

"Wilderness" gradually entered the North American language of conservation in the nineteenth century after the end of the frontier exploration, especially promoted by the hunting community. It developed as an aesthetic and ethical concept related to the protection of pristine nature in the face of galloping technological progress and rapid disappearance of natural environments. Thus wilderness became synonymous with freedom, natural beauty, sanctuary and retreat from everything that was perceived as overwhelming in the modern lifestyle (Nash 2001).

Some have argued that past landscape modifications by human populations and pervasive human impacts across scales make the idea of wilderness inconsequential (Heckenberger et al. 2003). Wilderness also attracted considerable controversy in North America, particularly raising questions relating to equity and the rights of humans living in, or next to, areas allocated to wilderness protection (Nash 2001). The same issues were raised on all other continents that were colonized by European settlers. The establishment of protected parks and hunting reserves in South Africa was accompanied by the relocation of native populations and social strife (Carruthers 1995). Australia has also experienced some controversy surrounding the definition of wilderness and its disconnection from the culture and lifestyle of aboriginal populations (Mackey et al. 1998).

Such developments gave "wilderness" the impetus to evolve towards a more relevant concept for the twenty-first century, incorporating both human dimensions and needs as well as new research results from areas such as paleoecology or climate science (Gillson and Willis 2004). A science-based understanding of the human influence on ecosystems informs presently one of the main current conservation approaches (Brooks et al. 2006; Kalamandeen and Gillson 2007). In this context, wilderness represents one extreme of the gradient of human presence and impact across the landscape. While still retaining an aesthetical element and an existence value among growing numbers of enthusiasts in the Western industrialized countries, wilderness also refers to the biophysical reality of natural processes, ecological communities, and the resulting ecosystems that develop in the absence of human management. Therefore, wilderness is of major importance both for research and management in the areas of ecosystem services (ES) (Naidoo et al. 2008, see Chap. 3), biodiversity conservation (Watson et al. 2009), and the establishment of ecosystem baselines (Vitousek et al. 2000).

Appreciation of European wilderness has had a different path from that on other continents due to the long history of human occupation, agriculture and landscape management. Many of the species that used to dominate the landscape in the distant past have been hunted to extinction or have been driven away from the most favourable habitats (Barnosky 2008, see Chaps. 4, 8) and natural vegetation cover has been cut or burnt down to make space for farmland. Thus both laymen and naturalists have come to regard and appreciate this new state as the natural biodiversity of the continent. As a result of a shifting baseline syndrome, traditional agricultural landscapes have become the benchmark against which biodiversity change was measured (Papworth et al. 2009). However, a growing movement in Europe advocates now for wilderness protection and recognition, and policy steps have been taken in this direction, including a resolution of the European Parliament on wilderness in Europe (Martin et al. 2008; European Parliament 2009). Research has also been undertaken in order to identify and map wilderness on the continent (Fritz et al. 2000; Carver 2010). In this favourable context, rewilding of abandoned farmland can gain momentum as a way of expanding the areas that provide both increased opportunities for wilderness experience and more extensive self-regulating and self-sustaining ecosystems (Rey Benayas et al. 2007; Munroe et al. 2013, see Chaps. 1, 11).

Considering the diversity of possible definitions, we approach wilderness in this chapter from several points of view. In the next section we review the literature on wilderness mapping and to identify some of the most important ecological and aesthetical aspects of wilderness in Europe. We then map and discuss the spatial agreement between wilderness based on (a) human access from roads and settlements, (b) impact of artificial light, (c) deviation from potential natural vegetation, and (d) proportion of primary productivity harvested by humans, as metrics of wilderness value over space. We further explore the health of trophic chains by looking at megafauna species and their spatial concurrence with wilderness. Megafauna such as the large herbivores, apex predators and birds of prey have an important role in maintaining and returning ecosystems to a higher naturalness state through establishment of natural trophic cascades (see Chaps. 4, 5, and 8). As such we also map the distribution of high body mass species across Europe and discuss the overlaps with high-wilderness quality and farmland abandonment areas. We then explore the possible spatial and temporal dynamics of wilderness in Europe over the next few decades in the context of farmland abandonment and rewilding. We examine how aspects of wilderness could increase due to agricultural abandonment and we suggest means to maximize the potential success of rewilding efforts.

2.2 Measuring and Mapping Wilderness—A Brief Review of Metrics and Methods

Wilderness has been mapped and analysed across scales, from global to local level. The methodologies generally make use of available spatial data on human infrastructures, land cover, area size of ecologically intact regions, etc. as proxies for wilderness quality, but also employ expert knowledge on degree of naturalness and ecosystem modification. Despite the obvious challenges of mapping a multidimensional concept such as wilderness, studies using relevant indicators at a similar extent and resolution offer highly congruent results, likely because they share a common perception of the attributes and values of wilderness.

At the global level, Mittermeier et al. (2003) used a combination of human population density, intactness, and area size of the intact areas to define wilderness areas. Much of their assessment was based on literature and expert opinions. The wilderness areas identified coincided with the areas of the lowest human footprint identified by Sanderson et al. (2002) although the two studies used largely different metrics. The map of the human footprint at the global level used human population density, the transformation of land through the building of settlements, roads and railroads, and measures of human access. Power infrastructures were also quantified, using satellite night maps (Sanderson et al. 2002). Despite data limitations, these global studies reveal a fairly consistent big picture of the overall pattern and magnitude of human impact on the biosphere, both for terrestrial and marine ecosystems (Halpern et al. 2008).

In Australia, the Heritage Commission's National Wilderness Inventory used four metrics for defining wilderness: remoteness from settlements, remoteness from access, biophysical naturalness and apparent naturalness (Lesslie et al. 1995). In this case, thresholds were defined for minimum levels of these metrics that would characterize wilderness. Other approaches emphasize a wilderness continuum across the landscape (Fritz et al. 2000). Building on the Heritage Commission's National Wilderness Inventory research, Carver et al. (2002) added remoteness from national population centres and altitude in order to map wilderness in the United Kingdom.

Remoteness from national population centres was a measure of the accessibility to the whole British population in addition to the accessibility to the local population in the calculation of wilderness. The authors used multicriteria evaluation (MCE) and explored public perceptions of wilderness through the use of interactive tools by allowing the user to change the weights of the wilderness metrics. As expected, resulting wilderness maps were not radically different, but allowed for insights on what affects the perceptions of wilderness (Carver et al. 2002). This approach was further detailed at the level of the Cairngorms National Park, and the Loch Lomond and The Trossachs National Park in Scotland (Carver et al. 2012) at a resolution of 20 m and later expanded to cover the whole of Scotland in a study by the Scottish Natural Heritage (Scottish Natural Heritage 2012).

At lower spatial extents the indicators of wilderness and human footprint remain the same but higher quality data are usually available making the mapping and modelling process more reliable and accurate. For example, Woolmer et al. (2008) rescaled the human footprint methodology of the Sanderson et al (2002) for the area of approximately 300,000 km^2 of the Northern Appalachian ecoregion. They used ten datasets compiled from several sources: population density, dwelling density, urban areas, roads, rail, land cover, large dams, watersheds, mine sites, utility corridors for the electrical power infrastructure. The general patterns of human footprint were maintained when comparing the map based on 90 m^2 resolution data at ecoregional scale with the map derived from the global analysis of Sanderson et al (2002) conducted with 1 km^2 resolution data. However, the Spearman rank correlation coefficients between the two sets of human footprint data steadily decreased with the scale, reaching 0.41 ($p < 0.001$) at 0.1% of the Northern Appalachian ecoregion. The difference in the human footprint scores is that the ecoregion calculation compared with the global calculation leads to a reduction in the area with low levels of human footprint (46% ecoregion extent vs. 59% global extent) and an increasing of the area with moderate or high levels of human footprint (34% ecoregion extent vs. 21% global extent), evening out more the distribution of human footprint scores. A key finding was also that three parameters models add the most information to the calculation of human footprint while the model incorporating human settlements, roads and land-use was the best approximating model from all combinations of the ten datasets considered.

In Europe, an increased wilderness momentum has led to efforts by different actors to protect wilderness and advance a progressive wilderness research agenda (Jones-Walters and Čivić 2010). A continental level map of wilderness continuum has been produced using population density, road and rail density, linear distance from the nearest road and railway line, naturalness of land cover and terrain ruggedness (Carver 2010). This analysis identified wilderness areas concentrated in the Scandinavian Peninsula and the mountainous regions of Europe, revealing a strong positive altitudinal and latitudinal relationship. The same pattern was maintained even if terrain ruggedness was eliminated from the calculation. Beside the Scandinavian mountains and arctic areas, the Pyrenees, The Eastern Mediterranean islands, the Alps, the British Isles, the south-eastern Europe and the Carpathians also had significant areas of wilderness (Carver 2010) but one has to temper this with the knowledge that the current spatial data often misses historical information on local land use management such as past deforestation, drainage and grazing by domestic livestock. Currently, the wilderness mapping is being updated through the project of the European Wilderness Registry, which will record the most important wild sites, thus facilitating priority setting for protection.

2.3   Wilderness Metrics

The set of metrics used in the wilderness mapping literature can be divided into two major dimensions of defining wilderness: the subjective or perceived wilderness experience and ecological intactness. Most wilderness metrics attempt to describe both aspects. For example, the presence of roads and human settlements indicate both easiness of access, visual impact, and the ecological impact of these infrastructures. Yet some indicators address the two dimensions separately as it is the case with apparent naturalness and biophysical naturalness (Lesslie et al. 1988). For the purposes of this chapter, we chose a series of four metrics: two that describe both the subjective human experience of wilderness and the ecological impact, and two that have mainly an ecological dimension. The metrics used here quantify human impact thus wilderness increases with the decrease of the metrics.

Remoteness from roads and human settlements is an important dimension in the feeling of solitude intrinsic to the wilderness experience. However, roads and other human access infrastructure have also a strong impact on wild populations and ecosystems. The most obvious impact is road mortality, shown to affect mammals (Philcox et al. 1999; Seiler 2005; Grilo et al. 2009), birds (Orlowski 2005), reptiles (Iosif et al. 2013) and amphibians (Patrick et al. 2012). But impacts of roads, traffic and human access can be much more profound, affecting population and community structure (Habib et al. 2007), trophic interactions (Kristan III and Boarman 2003; Whittington et al. 2011), ecosystem functioning and structure (Christensen et al. 1996; Hansen et al. 2005; Rentch et al. 2005), and environmental conditions through high pollution levels (Hatt et al. 2004). Roads can favour the expansion of invasive species (Jodoin et al. 2008; Vicente et al. 2010), and of exotic and human-favoured predators (Alterio et al. 1998). They also expose forest habitats to edge effects (Tabarelli et al. 2004). These ecological impacts of roads and human settlements alter a range of ecological conditions compared with the context that would exist without these human infrastructures. Here we evaluate human access from roads and settlements by calculating the cost distance to paved roads and settlements according to the Naismith's rule which assumes differentiated relative traveling times depending on terrain, land cover, and river networks (Carver and Fritz 1999). We extracted the data on paved roads from the Eurogeographics Road database and the Open Street Map database, land use data from Corine Land Cover 2000 and 2006, and terrain ruggedness data from the Shuttle Radar Topography Mission (SRTM) at 1 km resolution. The range of the human access score values is expressed from 0 to 1. In Europe, the mountainous areas, the Iberian Peninsula, the Balkans, Scotland, and Scandinavia are the least accessible regions and the least impacted by roads and settlements (Fig. 2.1a).

Artificial night light has a similar dimension in the definition of wilderness. Light pollution has been decried for its impact on the visibility of the natural night sky (Cinzano et al. 2000), diminishing the night wilderness experience. But artificial light has also strong ecological impacts (Longcore and Rich 2004; Navara and Nelson 2007; Hölker et al. 2010b; Gaston et al. 2013), affecting invertebrates (Davies et al. 2012, see Chap. 6), fish (Becker et al. 2013), mammals (Boldogh et al. 2007) and bird populations (Montevecchi et al. 2006). Direct mortality (Hölker et al. 2010b), impacts on trophic relations and community structure (Perkin et al. 2011), disruption of migratory routes (Gauthreaux Jr et al. 2006) by night light lead to profound modifications of ecosystems functions (Hölker et al. 2010a). Nocturnal species such as bats and moths (see also Chap. 6) receive the brunt of the impact. We assess the impact of artificial light on ecosystems and wilderness experience by using the satellite data of the upwards emitted and reflected artificial light with a spectral range of 0.5–0.9 µm in Europe from the Visible Infrared Imaging Radiometer Suite (VIIRS) of the Soumi National Polar-orbiting Partnership (SNPP) for the year 2012 (NOAA National Geophysical Data Center 2012) with a resolution of 15 arc sec (approximately 450 m). We apply a kernel function to distribute the impact over a radius of approximately 10 km (Fig. 2.1b) as a conservative approximation meant to cover the night glow effects reported in the literature (Kyba et al. 2011) along with the direct ecological impacts (Longcore and Rich 2004). In each pixel, the light impact score is the sum of all the impact scores from the surrounding light sources and it represents a relative measure aimed at encompassing both the ecological aspect and the impact on the subjective wilderness experience (Fig. 2.1b).

The last two metrics that we consider here are qualitative and quantitative measures of the human modification of ecosystems and thus they convey mainly, although not exclusively, an ecological significance. Anthropogenic change of natural habitat is one of the major drivers of biodiversity loss (Pereira et al. 2010) and it has been studied extensively for a large range of taxa (Bolliger et al. 2007). The most conspicuous element of habitat loss is the change in vegetation, and intact vegetation cover has been used before as a wilderness indicator (Bryant et al. 1997). Human changes in vegetation tips the balance in favour of species benefiting from human presence and impacts habitat-sensitive ones (Leu et al. 2008). Therefore we use here the deviation from potential natural vegetation (dPNV) as a qualitative measure of the human impact on the landscape. We used the potential natural vegetation (PNV) classes of the map developed by Bohn et al. (2000). We calculate the similarity of current land cover to PNV by estimating the probability that the CORINE 2000 land cover class in any one location in Europe belongs to the local PNV type (Bohn et al. 2000). The probability of agreement was classified in four classes with different scores: assumed = 1, most probable = 0.75, probable = 0.5 and possible = 0.1. The resulting map was combined with the grazing density data from Food and Agriculture Organization, which was previously linear transformed to a scale from 0 to 1, where 1 represents a density of 20 heads/km^2 or more. We used

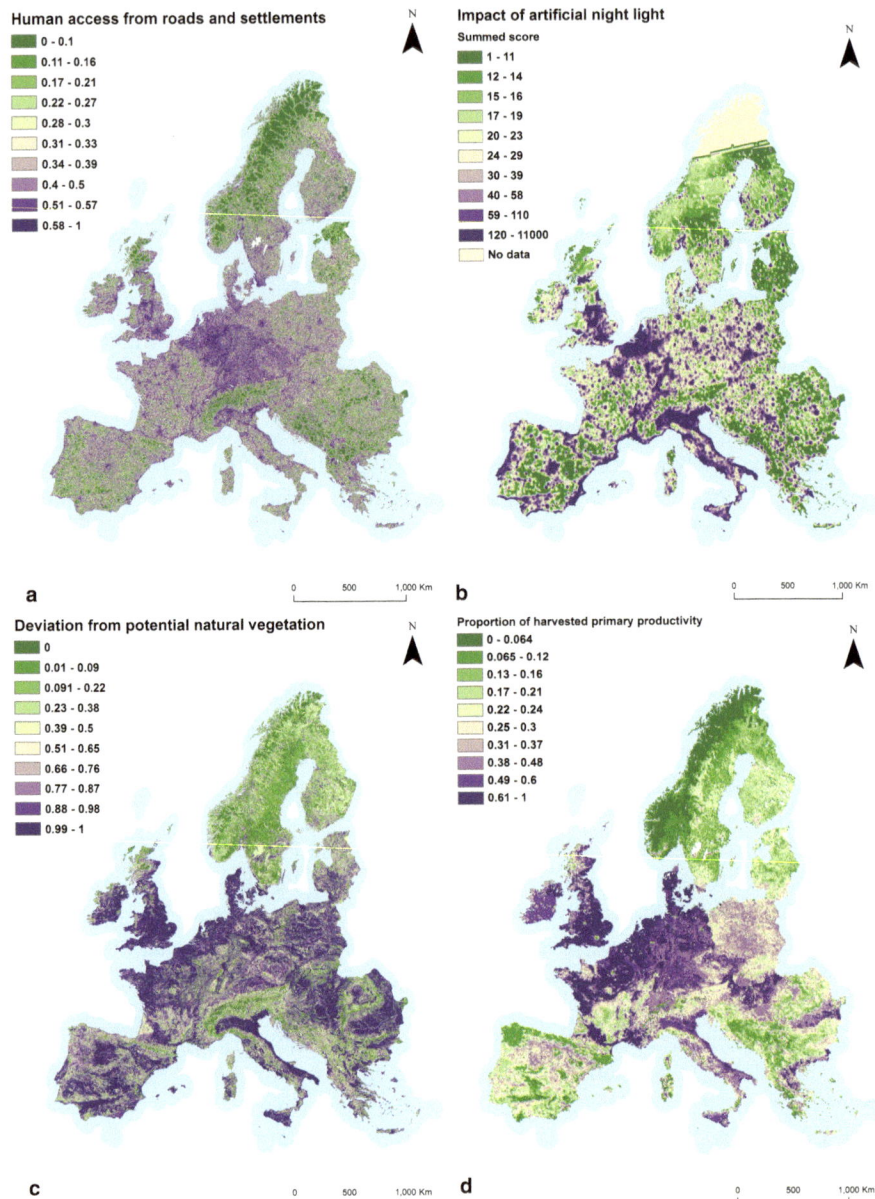

Fig. 2.1 Wilderness areas according to four metrics. **a** Access from roads and human settlements. **b** Artificial night light. **c** Deviation from potential natural vegetation. **d** Proportion of harvested primary productivity out of the potential primary productivity. Wilderness value increases with the decrease of the metrics

grazing density to account for human transformations in semi-natural grasslands. We expressed the dPNV value by subtracting from 1 the score calculated according to the described methodology. (Fig. 2.1c).

Through agriculture, hunting, fishing and forestry, humans are removing significant quantities of biomass from the ecosystems. Primary productivity (PP) is the foundation of trophic networks and it influences the structure and functions of ecosystems in a domino effect across trophic levels (Haberl et al. 2004). Humans have reduced drastically the PP available to other species and this has changed the composition of the ecological communities (Barnosky 2008; Pereira et al. 2012). We map the proportion of human harvested PP out of the total potential PP in Europe as another indicator of wilderness and using the data analysed in Haberl et al. (2007). We calculated the harvested PP by extracting net PP remaining in ecosystems after harvest from the net PP of the actual vegetation. We then calculated the proportion of harvested PP by dividing net harvested PP by net PP of the potential vegetation. The data are calculated based on country-level statistics of the Food and Agriculture Organization (Haberl et al. 2007) while potential PP is estimated using the Lund-Potsdam-Jena dynamic global vegetation model (Sitch et al. 2003). Some abnormalities can be noticed in the harvested PP map which are due to the assumptions of the model and the FAO national level data. The map has to be interpreted with this limitation in mind (Fig. 2.1d).

The four resulting maps based on the selected metrics show a common pattern of high human footprint in the lowlands of central Europe (Fig. 2.1). The most unaltered values of all metrics occur in high mountainous areas and Scandinavia. But the differences at intermediate values of wilderness provide a key signal to what are the strongest determinants of human footprint at regional level in Europe. For example, although the dPNV is very low in almost all of Scandinavia (Fig. 2.1c), the proportion of harvested PP is comparatively higher, consistent with high forestry harvest in the Nordic countries (Fig. 2.1d). The reverse pattern is noticeable in the Iberian Peninsula where although the drier climate restricts high harvesting of PP, the current vegetation is quite far from PNV as measured in our map and consistent with the degradation of the Mediterranean habitats (Myers et al. 2000). In the same region, the significant differences between the inland and coastal values of the night light impact and human access (Fig. 2.1a and b) indicates the high difference between the human population densities inland compared with the coastal regions. These differences in the distribution of human populations are masked in the PNV score and harvested PP maps (Fig. 2.1c and d). The map of artificial light (Fig. 2.1b) also points out to a discrepancy in the relative wilderness values in East and South-East Europe compared with the dPNV score map for example (Fig. 2.1c). The lower economic activity in this area results in lower light impact although the level of vegetation change is very high (Doll et al. 2006).

The lowest wilderness areas in Europe have usually low scores for all the wilderness dimensions considered, and they represent mainly areas of high human densities and intense economic activity. Conversely, high wilderness areas are the wildest from all the points of view taken here. But the areas of intermediate wilderness values are strongly impacted by only one or two metrics with very low wilderness values. Especially dPNV and harvested PP have a farther reach, affecting even eco-systems where infrastructure and artificial light impacts are reduced. These indica-tors are connected with more extensive land-uses such as agriculture and forestry, and less with high human population densities and infrastructure.

The synergies and interactions between the different elements of our wilder-ness mapping emphasize even further their ecological significance. In areas of high habitat quality the road mortality can be higher in absolute terms because it affects more abundant populations (Patrick et al. 2012) while road lighting can increase the impact of the road itself on the local ecological communities by favouring certain types of predation (Rich and Longcore 2005) or providing additional perches for improved hunting efficiency of raptors such as kestrels (Sheffield et al. 2001).

2.4 Wilderness Conservation

The designation, coverage and implementation of protected areas and Natura 2000 sites vary widely across European countries. However, looking at the continental map, we discern some regional patterns in wilderness protection. Many mountain-ous areas in the Pyrenees, the Apennines, the Massif Central and the Carpathians are covered by Natura 2000 sites and, to a lesser extent, by nationally designated pro-tected areas (Fig. 2.2) (European Environment Agency 2012a, b). Large protected areas included both in the Natura 2000 network and in the national networks protect the Scandinavian mountains. As already pointed out in the literature (Gaston et al. 2008), many of the designated areas overlap because countries have co-designated under Natura 2000 and their own national systems. However, important differences between the two protected areas systems can also be noticed (Fig. 2.2). For ex-ample, the Iberian Peninsula and South-Eastern Europe seem to have a much larger area under protection by the Natura 2000 network than from nationally designated protected areas. Conservation seems to have benefitted in these areas from a push from the European conservation policies (European Council 1979, 1992). Mean-while, Germany and France have smaller and fewer terrestrial protected areas under the Natura 2000 network than under the national network.

It has been suggested in the literature that the designation of protected areas has been done opportunistically and thus that they are more likely to cover low produc-tivity, high altitude, wilderness areas (Pressey et al. 1993; Margules and Pressey 2000). Although largely lacking continental coordination, Natura 2000 network has some features common with systematic conservation planning and aims to protect species and habitats threatened at continental level (Gaston et al. 2008). Surpris-ingly however, the terrestrial Natura 2000 sites have a lower continental average proportion of harvested PP than nationally designated protected areas: 26.7 % for Natura 2000 sites against 34.3 % for the nationally designated protected areas. The continental average values for the impact of artificial night light in Natura 2000 sites is 38 while in nationally designated protected areas network is 31, showing the same pattern as in the case of harvested PP. However, we have to keep in mind that

Nationally designated protected areas

Natura 2000 sites

N

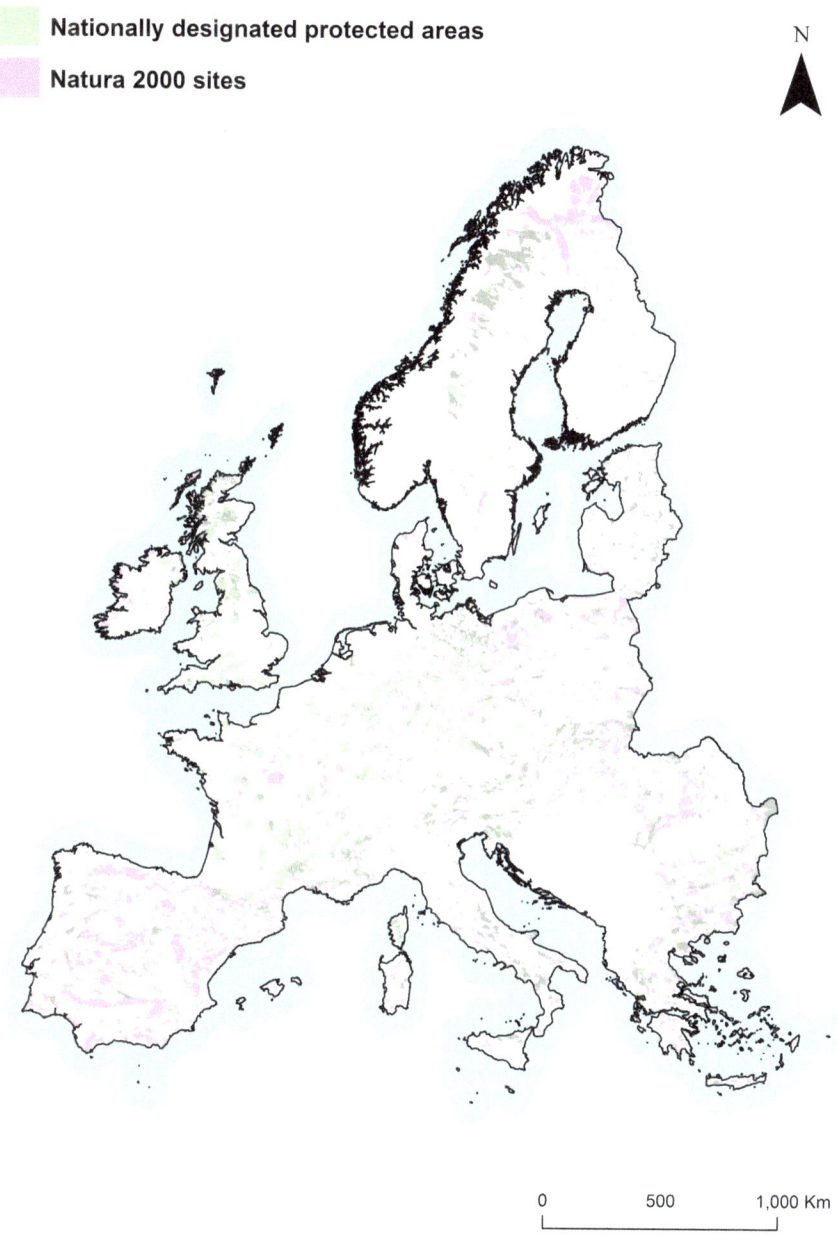

| 0 | 500 | 1,000 Km |

Fig. 2.2 Protected territory in Europe under the Natura 2000 network and nationally designated protected areas

there are big regional differences between the patterns of wilderness in protected areas in Europe. For instance, analysis concentrating on Germany as a case study demonstrated that the Natura 2000 areas in Western Germany largely fail to protect the roadless and the low-traffic areas, whereas in former East Germany a better congruence was achieved (Selva et al. 2011).

Indicative of higher resource availability, we verified that higher species richness of megafauna species coincides with high wilderness. We selected the mammals with an adult bodyweight of an average of 10 kg or more (Jones et al. 2009) from the data of the Atlas of European Mammals (Mitchell-Jones et al. 1999). These include species of large herbivores and apex predators such as the wolf (*Canis lupus*) and the lynx (*Lynx lynx*). We also selected the bird species with an adult bodyweight of an average of 5 kg or more (Myers et al. 2013; Tacutu et al. 2013) using data from the atlas of the European Bird Census Council (Hagemeijer and Blair 1997). These species include several birds of prey as well as other species such as the great white pelican (*Pelecanus onocrotalus*) or the great bustard (*Ardeotis nigriceps*). In the end, we obtained a megafauna list of 30 mammal species and 13 bird species distributed in a grid of 50×50 km^2 covering the European territory. At a visual examination, the highest species richness areas in terms of megafauna coincide with high wilderness areas in Europe such as the Carpathians, the Apennines and the Pyrenees (Fig. 2.3). We calculated rank correlations between the megafauna species richness and average values per grid cell of the four wilderness metrics. The results suggest that wilderness and megfauna populations spatially coincide in Europe ($\rho = 0.18$, $p < 0.0001$ for access from roads and settlements, $\rho = -0.28$, $p < 0.0001$ for light impact, $\rho = 0.34$, $p < 0.0001$ for dPNV score, $\rho = -0.26$, $p < 0.0001$ for harvested PP). There are several mechanisms that could underlie this pattern such as the direct persecution of carnivores and birds of prey countered by conservation programs in areas of lowest social conflict (Valkama et al. 2005; Enserink and Vogel 2006). This pattern could also be related to a phylogenetic bias determined by the strong predominance of a few bird and mammal orders in our selection which could be limited to certain habitats only based on their common evolutionary history. We also did not consider the possible spatial autocorrelation in our datasets. However, from the perspective of abandonment, the spatial concurrence between megafauna species richness and high wilderness is important because it means that abandoned farmland closer to high wilderness areas will have a better chance of being repopulated by these species. This will lead to a quicker recovery of trophic networks and natural ecological processes.

2.5 Farmland Abandonment as Opportunity for Wilderness Expansion

Farmland abandonment in Europe is a result of the economic and social changes at national, continental and global levels. Abandonment happens especially in areas where land productivity is not sufficiently high to sustain an adequate income for

Farmland abandonment

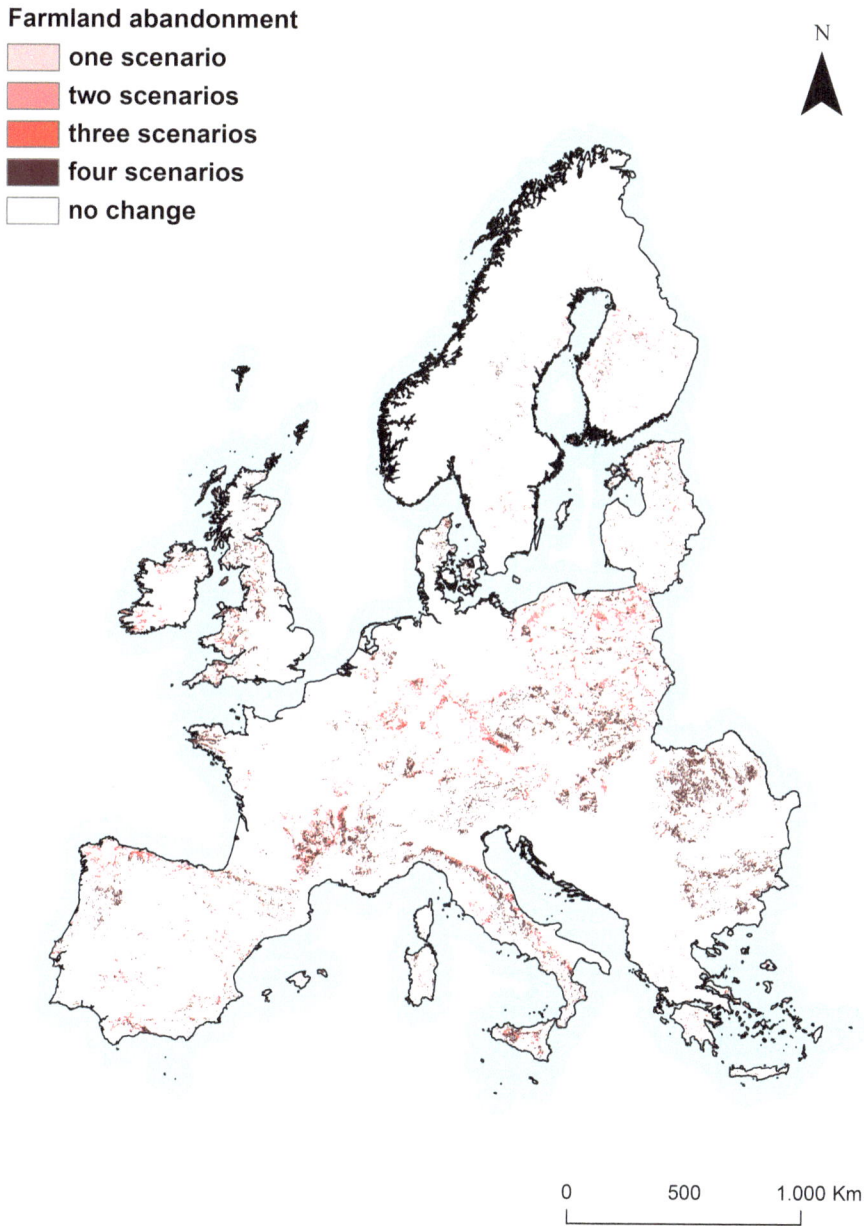

Fig. 2.3 Farmland abandonment in Europe projected for the year 2040 by the Dyna-CLUE model based on four VOLANTE scenarios. We indicate in how many of the four scenarios land abandonment is found significant across the continent

farmers, even with the support of subsidies (Rey Benayas et al. 2007, see Chap. 1). These land-use changes raise challenges in terms of lifestyles, social structure and biodiversity (Munroe et al. 2013). Thus, predicting these changes has received considerable importance in recent research. We map the areas in Europe where farmland abandonment is projected to take place based on the Dyna-CLUE model (Verburg and Overmars 2009) (Fig. 2.3). For the projections of the social and economic drivers driving farmland abandonment, we used four VOLANTE scenarios describing different development paths towards the year 2040 (Paterson et al. 2012). These scenarios are loosely based on the Special Report on Emission Scenarios (SRES) of the Inter-governmental Panel on Climate Change (Nakicenovic et al. 2000) and they cover the range of socio-economic conditions across the axes of regionalization versus globalization, and willingness versus reluctance against sustainable lifestyle changes at the societal level. We indicate here in how many of the four scenarios land abandonment is found significant across Europe (Fig. 2.3).

How will farmland abandonment affect wilderness value? The answer to this question depends on where farmland abandonment takes place. Many areas of abandonment can be found around mountainous regions such as the Apennines, the Massif Central, the Carpathians, the Balkans, areas of higher altitude and lower productivity that have already experienced abandonment in the past decades (Fig. 2.3). These areas have a low density of human population and a low level of infrastructure development. As the human density will decrease even more, the use of artificial light will decrease as well, but the physical infrastructures will withstand for longer than the outmigration of people albeit with lower intensity of use. Spurred by the already existing infrastructure, many abandonment areas might also see a surge in tourism, biofuels cultivation and renewable energy industries, replacing the agricultural activities (Laiolo and Tella 2006).

From an ecological point of view, farmland abandonment will directly lead to a decrease in harvested PP as grazing and cultivation are projected to drop. This will increase the resources available to wild populations and ecosystems, and vegetation cover will evolve towards a more natural state (Rey Benayas et al. 2007). Previous studies have showed that increased availability of biomass and reduced presence of humans lead to growing numbers of wild herbivores in south Asia (Madhusudan 2004). The recovery of ecosystems to a wilderness state depends on rebuilding natural trophic cascades and networks that are both resilient to natural disturbances and able to sustain key ecosystem functions. In these networks, megafauna and apex predators have a fundamental role, especially in the depleted conditions of the current European biota (Schmitz 2006; Sekercioglu 2006; Johnson 2009; Ritchie and Johnson 2009). For the natural recovery of ecosystems and the return of these species, the presence of source populations is paramount and adjacency to existing core wilderness areas will be a key driver (see Chaps. 4, 8).

We explore the chances for a natural recovery of European fauna by mapping the distribution of megafauna (Fig. 2.4). The results are encouraging for many areas of future agricultural abandonment: megafauna richness is high in the adjacent areas and many wild populations have already begun to recover, especially in the case of mammals (Enserink and Vogel 2006). In the case of birds, the literature reports significant changes in the community patterns due to abandonment, especially negative effects on populations of farmland birds with narrow habitat preferences (Sirami et al. 2008). The correlation between the number of mammal species and the percentage of projected abandoned area in a grid cell is $\rho = 0.14$ ($p < 0.001$) whereas for bird species it is negative at $\rho = -0.15$ ($p < 0.001$). Thus megafauna mammal species might be in a better position to take advantage of the new resources and space made available by farmland abandonment. We did not consider here the possible spatial autocorrelation of the data because we were interested only in the spatial coincidence between abandonment and megafauna.

However, some of the future abandoned areas have been affected by invasive species, fire suppression practices, and missing trophic links during thousands of years of human use (Proença et al. 2010; Wehn et al. 2011). Thus abandonment may not be sufficient to return these areas to a vegetation close to PNV in a short term without management actions (see Chap. 8). But even in these areas the abandonment will have immediate positive effects on wildlife by reducing human disturbance, increasing landscape connectivity, and releasing ecological processes from human control and thus increasing the wilderness value of the land (see Chap. 1).

Aplet et al. (2000) describe the two dimensional space defined by the axes of freedom and naturalness as a framework for wilderness management. Freedom is understood as the absence of human control over ecological processes (i.e self-willed) while naturalness is the degree to which ecosystems are close to an accepted ecological benchmark. Such a framework is readily usable for mapping the trade-offs related to human management in areas affected by invasive and exotic species, thus increasing naturalness but decreasing freedom (Landres et al. 2000; Sydoriak et al. 2000), but also the current views on rewilding as some advocate for serious management commitments in order to achieve a certain perception of wilderness (Donlan et al. 2006). However, we consider that the ultimate aim of rewilding is not to recreate some image of pre-human ecosystems, but to facilitate new, self-regulating systems that appear naturally out of the current conditions. A realistic expectation is that in the absence of human management, the new rewilded areas will form novel ecosystems that share elements with the pre-human past but also integrate current factors. Minimum human management and this new wilderness of natural and self-sustaining ecosystems should be the goal of rewilding.

2.6   Conclusions

Wilderness in Europe has been pushed into the high altitude areas of the mountain ranges and into the high latitude areas of Scandinavia. The metrics we use here agree on the general patterns of European wilderness but regional differences between our metrics emphasize the different factors that affect wilderness values regionally and locally (Fig. 2.1). New opportunities for wilderness expansion have appeared in Europe

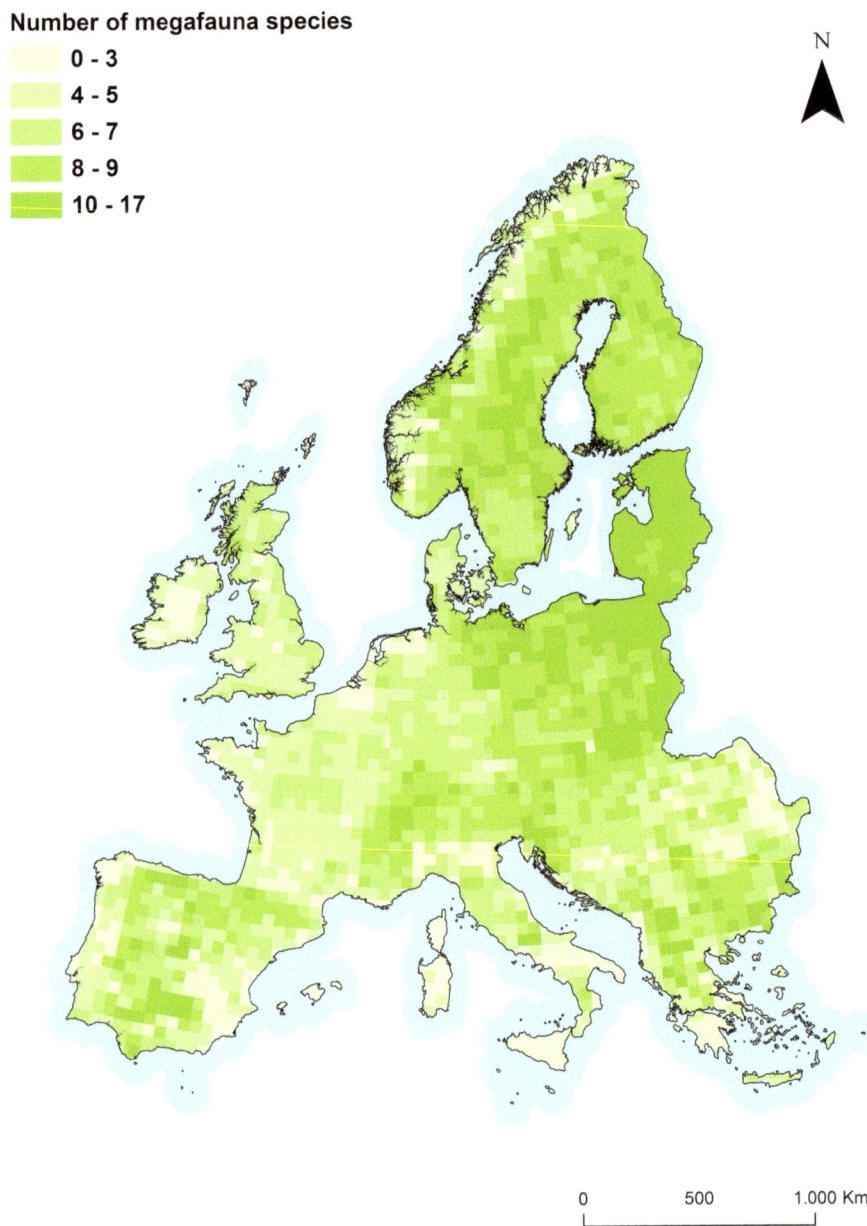

Number of megafauna species

- 0 - 3
- 4 - 5
- 6 - 7
- 8 - 9
- 10 - 17

Fig. 2.4 Species richness of European megafauna. We calculate it as the number of species of mammals with adult body mass equal or higher than 10 kg and birds with adult body mass equal or higher than 5 kg in each grid cell

due to farmland abandonment and a decrease of human presence can lead to a drop in several human footprint indicators and a recovery of natural trophic networks. Although management trade-offs have to be made in some places between intervening for a faster recovery and stepping back for an unrestrained adaptation of ecosystems, we favour an approach of minimum intervention and self-regulating ecosystems.

The next few decades are crucial for how wilderness will evolve in Europe. New research is necessary on how different dimensions of wilderness will change as a result of land use changes and what will be the effects on ecosystems and wildlife. Moreover, research is need on how to restore not only ecosystems but also the collective memory to encompass what wilderness may have been like, what it is and what it may or should be. This would help consolidate the crucial link between research on one hand, and management and policy on the other, an area that still requires substantial work (see Chap. 11). Challenges remain in bringing together rewilding views, and negotiating diverging social and economic interests. A focus on the benefits of natural ecosystem for the society at large can ease the tensions between different stakeholders in continental policy-making. From a global perspective, Europe will continue to be at the lower end of the wilderness continuum (Mittermeier et al. 2003) but favourable opportunities are arising at continental level to improve ecosystem functions and we should seize them wisely.

Acknowledgments PV acknowledges funding from FP7 project VOLANTE and OPERAs under which the scenario runs of land abandonment were performed. SC had a doctoral grant from Fundação para a Ciência e a Tecnologia (FCT) (SFRH/BD/80230/2011) until September 30, 2013. We thank Societas Europaea Mammalogica and Tony Mitchell-Jones for sharing and allowing the use of the data on European mammals and Michiel Van Eupen for providing assistance with the Wilderness Register data. We also thank Carlos Teixeira for suggesting sources on bird life traits and Guy Pe'er, Christoph Plutzar and Laetitia M. Navarro for comments on earlier drafts of this chapter.

References

Alterio, N., Moller, H., & Ratz, H. (1998). Movements and habitat use of feral house cats Felis catus, stoats Mustela erminea and ferrets Mustela furo, in grassland surrounding Yellow-eyed penguin Megadyptes antipodes breeding areas in spring. *Biological Conservation, 83,* 187–194.

Aplet, G., Thomson, J., & Wilbert, M. (2000). Indicators of wildness: Using attributes of the land to assess the context of wilderness. Proceedings: Wilderness Science in a Time of Change. Ogden (UT): USDA Forest Service, Rocky Mountain Research Station. Proc. RMRS-P-15

Barnosky, A. D. (2008). Megafauna biomass tradeoff as a driver of quaternary and future extinctions. *Proceedings of the National Academy of Sciences, 105,* 11543–11548. doi:10.1073/pnas.0801918105.

Becker, A., Whitfield, A. K., Cowley, P. D., et al. (2013). Potential effects of artificial light associated with anthropogenic infrastructure on the abundance and foraging behaviour of estuary-associated fishes. *Journal of Applied Ecology, 50,* 43–50.

Bohn, U., Gollub, G., Hettwer, C., et al. (2000). *Karte der natürlichen Vegetation Europas, Maßstab 1: 2,500,000.[Map of the Natural vegetation of Europe. Scale 1: 2,500,000].* Bonn: Bundesamt für Naturschutz.

Boldogh, S., Dobrosi, D., & Samu, P. (2007). The effects of the illumination of buildings on house-dwelling bats and its conservation consequences. *Acta Chiropterologica, 9,* 527–534.

Bolliger, J., Kienast, F., Soliva, R., & Rutherford, G. (2007). Spatial sensitivity of species habitat patterns to scenarios of land use change (Switzerland). *Landscape Ecology, 22,* 773–789.

Brooks, T. M., Mittermeier, R. A., da Fonseca G. A., et al. (2006). Global biodiversity conservation priorities. *Science, 313,* 58.

Bryant, D., Nielsen, D., & Tangley, L. (1997). *Last frontier forests: Ecosystems and economies on the edge.* Washington, D. C.: World Resources Institute, Forest Frontiers Initiative.

Carruthers, J. (1995) *The Kruger National Park: A social and political history.* Pietermaritzburg: University of Natal Press.

Carver, S. (2010). Mountains and wilderness. European Environment Agency (2010) Europe's ecological backbone: Recognising the true value of our mountains. European Environment Agency, Copenhagen, pp. 192–201.

Carver S., Evans AJ., Fritz S. (2002). Wilderness attribute mapping in the United Kingdom. *International Journal of Wilderness 8,* 24–29.

Carver, S., & Fritz, S. (1999). Mapping remote areas using GIS. Landscape character: Perspectives on management and change Natural Heritage of Scotland Series, HMSO, pp. 112–126.

Carver, S., Comber, A., McMorran, R., & Nutter, S. (2012). A GIS model for mapping spatial patterns and distribution of wild land in Scotland. *Landscape and Urban Planning, 104,* 395–409.

Christensen, D. L., Herwig, B. R., Schindler, D. E., & Carpenter, S. R. (1996). Impacts of lakeshore residential development on coarse woody debris in north temperate lakes. *Ecological Applications, 6,* 1143–1149.

Cinzano, P., Falchi, F., Elvidge, C. D., & Baugh, K. E. (2000). The artificial night sky brightness mapped from DMSP satellite Operational Linescan System measurements. *Monthly Notices of the Royal Astronomical Society, 318,* 641–657.

Cole, D. N., & Landres, P. B. (1996). Threats to wilderness ecosystems: Impacts and research needs. *Ecological Applications, 6,* 168–184.

Davies, T. W., Bennie, J., & Gaston, K. J. (2012). Street lighting changes the composition of invertebrate communities. *Biology Letters, 8,* 764–767.

Doll, C. N., Muller, J.-P., & Morley, J. G. (2006). Mapping regional economic activity from night-time light satellite imagery. *Ecological Economics, 57,* 75–92.

Donlan, C. J., Berger, J., Bock, C. E., et al. (2006). Pleistocene rewilding: An optimistic agenda for twenty-first century conservation. *The American Naturalist, 168,* 660–681.

Enserink, M., & Vogel, G. (2006) The carnivore comeback. *Science, 314,* 7476–7749.

European Council. (1979). Council Directive 79/409/EEC on the conservation of wild birds.

European Council. (1992). EU Habitats Directive (92/43/EEC). Consolidated Text. Office for Official Publication of The European Union. CONSLEG: 1992LOO43-01.05-2004.

European Environment Agency. (2012a). Natura 2000 data-the European network of protected sites. http://www.eea.europa.eu/data-and-maps/data/natura-3. Accessed 5 July 2013.

European Environment Agency. (2012b). Nationally designated areas (CDDA). http://www.eea.europa.eu/data-and-maps/data/nationally-designated-areas-national-cdda-7. Accessed 5 July 2013.

European Parliament. (2009). Resolution on Wilderness in Europe.

Fritz, S., Carver, S., & See, L. (2000). New GIS approaches to wild land mapping in Europe. Proceedings of the Wilderness science in a time of change conference. USDA Forest Service, Missoula, Montana, pp. 120–127

Gaston, K. J., Jackson, S. F., Nagy, A., et al. (2008). Protected areas in Europe. *Annals of the New York Academy of Sciences, 1134,* 97–119.

Gaston, K. J., Bennie, J., Davies, T. W., & Hopkins, J. (2013). The ecological impacts of nighttime light pollution: A mechanistic appraisal. *Biological Reviews, 88,* 912–927.

Gauthreaux, S. A., Jr., Belser, C., Rich, C., & Longcore, T. (2006). Effects of artificial night lighting on migrating birds. In C. Rich & T. Longcore (Eds.) Ecological Consequences of Artificial Night Lighting (pp.67–93). Washington D.C.: Island Press.

Gillson, L., & Willis, K. J. (2004). As earth's testimonies tell: Wilderness conservation in a changing world. *Ecology Letters, 7,* 990–998.

Grilo, C., Bissonette, J. A., & Santos-Reis, M. (2009). Spatial-temporal patterns in Mediterranean carnivore road casualties: Consequences for mitigation. *Biological Conservation, 142,* 301–313.

Haberl, H., Schulz, N. B., Plutzar, C., et al. (2004). Human appropriation of net primary production and species diversity in agricultural landscapes. *Agriculture, Ecosystems & Environment, 102,* 213–218.

Haberl, H., Erb, K. H., Krausmann, F., et al. (2007). Quantifying and mapping the human appropriation of net primary production in earth's terrestrial ecosystems. *Proceedings of the National Academy of Sciences, 104,* 12942.

Habib, L., Bayne, E. M., & Boutin, S. (2007). Chronic industrial noise affects pairing success and age structure of ovenbirds Seiurus aurocapilla. *Journal of Applied Ecology, 44,* 176–184.

Hagemeijer, W. J., & Blair, M. J. (1997). *The EBCC atlas of European breeding birds: Their distribution and abundance.* London: T. & AD Poyser.

Halpern, B. S., Walbridge, S., Selkoe, K. A., et al. (2008). A global map of human impact on marine ecosystems. *Science, 319,* 948–952.

Hansen, A. J., Knight, R. L., Marzluff, J. M., et al. (2005). Effects of exurban development on biodiversity: Patterns, mechanisms, and research needs. *Ecological Applications, 15,* 1893–1905.

Hatt, B. E., Fletcher, T. D., Walsh, C. J., & Taylor, S. L. (2004). The influence of urban density and drainage infrastructure on the concentrations and loads of pollutants in small streams. *Environmental Management, 34,* 112–124.

Heckenberger, M. J., Kuikuro, A., Kuikuro, U. T., et al. (2003). Amazonia 1492: Pristine forest or cultural parkland? *Science, 301,* 1710–1714.

Hochtl, F., Lehringer, S., & Konold, W. (2005). "Wilderness": What it means when it becomes a reality-a case study from the southwestern Alps. *Landscape and Urban Planning, 70,* 85–95.

Hölker, F., Moss, T., Griefahn, B., et al. (2010a). The dark side of light: A transdisciplinary research agenda for light pollution policy. *Ecology and Society, 15*(4): 13.

Hölker, F., Wolter, C., Perkin, E. K., & Tockner, K. (2010b). Light pollution as a biodiversity threat. *Trends in ecology & evolution, 25,* 681–682.

Iosif, R., Rozylowicz, L., & Popescu, V. D. (2013). Modeling road mortality hotspots of Eastern Hermann's tortoise in Romania. *Amphibia-Reptilia, 34,* 163–172.

Jodoin, Y., Lavoie, C., Villeneuve, P., et al. (2008). Highways as corridors and habitats for the invasive common reed Phragmites australis in Quebec, Canada. *Journal of Applied Ecology, 45,* 459–466.

Johnson, C. N. (2009). Ecological consequences of Late Quaternary extinctions of megafauna. *Proceedings of the Royal Society B: Biological Sciences, 276,* 2509–2519.

Jones, K. E., Bielby, J., Cardillo, M., et al. (2009). PanTHERIA: A species-level database of life history, ecology, and geography of extant and recently extinct mammals: Ecological Archives E090-184. *Ecology, 90,* 2648–2648.

Jones-Walters, L., & Čivić, K. (2010). Wilderness and biodiversity. *Journal for Nature Conservation, 18,* 338–339.

Kalamandeen, M., & Gillson, L. (2007). Demything "wilderness": Implications for protected area designation and management. *Biodiversity and Conservation, 16,* 165–182.

Kristan, W. B. III, & Boarman, W. I. (2003). Spatial pattern of risk of common raven predation on desert tortoises. *Ecology, 84,* 2432–2443.

Kyba, C. C., Ruhtz, T., Fischer, J., & Hölker, F. (2011). Cloud coverage acts as an amplifier for ecological light pollution in urban ecosystems. *PLoS ONE, 6,* e17307.

Laiolo, P., & Tella, J. L. (2006). Fate of unproductive and unattractive habitats: Recent changes in Iberian steppes and their effects on endangered avifauna. *Environmental Conservation, 33,* 223–232.

Landres, P. B., Brunson, M. W., Merigliano, L., et al. (2000). Naturalness and wildness: The dilemma and irony of managing wilderness. *Proceedings RMRS-P-15Proceedings RMRSP-15, 5*, 377–381.

Lesslie, R. G., Mackey, B. G., & Preece, K. M. (1988). A computer-based method of wilderness evaluation. *Environmental Conservation, 15*, 225–232.

Lesslie, R. G., Maslen, A., Commission AH. (1995). National Wilderness Inventory Australia: Handbook of Procedures, Content, and Usage. Australian Government Pub. Service

Leu, M., Hanser, S. E., & Knick, S. T. (2008). The human footprint in the west: A large-scale analysis of anthropogenic impacts. *Ecological Applications, 18*, 1119–1139.

Longcore, T., & Rich, C. (2004). Ecological light pollution. *Frontiers in Ecology and the Environment, 2*, 191–198.

Mackey, B. G., Lesslie, R. G., Lindenmayer, D. B., et al. (1998). *The role of wilderness in nature conservation*. Australia: Canberra.

Madhusudan, M. D. (2004). Recovery of wild large herbivores following livestock decline in a tropical Indian wildlife reserve. *Journal of Applied Ecology, 41*, 858–869.

Margules, C. R., & Pressey, R. L. (2000). Systematic conservation planning. *Nature, 405*, 243–253.

Martin, V. G., Kormos, C. F., Zunino, F., et al. (2008). Wilderness Momentum in. Europe. *International Journal of Wilderness, 14*, 34–38.

Mitchell-Jones, A. J., Amori, G., Bogdanowicz, W., et al. (1999). *The atlas of European mammals*. London: Academic.

Mittermeier, R. A., Mittermeier, C. G., Brooks, T. M., et al. (2003). Wilderness and biodiversity conservation. *Proceedings of the National Academy of Sciences, 100*, 10309.

Montevecchi, W. A., Rich, C., & Longcore, T. (2006). Influences of artificial light on marine birds. In C. Rich & T. Longcore (Eds.) Ecological Consequences of Artificial Night Lighting (pp.94–113). Washington D.C.: Island Press

Munroe, D. K., van Berkel D. B., Verburg, P. H., & Olson, J. L. (2013). Alternative trajectories of land abandonment: Causes, consequences and research challenges. *Current Opinion in Environmental Sustainability, 5*, 471–476.

Myers, N., Mittermeier, R. A., Mittermeier, C. G., et al. (2000). Biodiversity hotspots for conservation priorities. *Nature, 403*, 853–858.

Myers, P., Espinosa, R., Parr, C. S., et al. (2013). The Animal Diversity Web (online). http://animaldiversity.org. Accesssed 10 July 2013.

Naidoo, R., Balmford, A., Costanza, R., et al. (2008). Global mapping of ecosystem services and conservation priorities. *Proceedings of the National Academy of Sciences, 105*, 9495.

Nakicenovic, N., Alcamo, J., Davis, G., et al. (2000). Special report on emissions scenarios: A special report of Working Group III of the Intergovernmental Panel on Climate Change. Pacific Northwest National Laboratory, Richland, WA (US), Environmental Molecular Sciences Laboratory (US).

Nash, R. (2001) *Wilderness and the American mind*. London: Yale Nota Bene.

Navara, K. J., & Nelson, R. J. (2007). The dark side of light at night: Physiological, epidemiological, and ecological consequences. *Journal of pineal research, 43*, 215–224.

NOAA National Geophysical Data Center EOG. (2012). VIIRS Nighttime Lights—2012.

Orlowski, G. (2005). Factors affecting road mortality of the Barn Swallows Hirundo rustica in farmland. *Acta Ornithologica, 40*, 117–125.

Papworth, S. K., Rist, J., Coad, L., & Milner-Gulland, E. J. (2009). Evidence for shifting baseline syndrome in conservation. *Conservation Letters, 2*, 93–100.

Paterson, J., Metzger, M., & Walz, A. (2012). Deliverable No: 9.1—The VOLANTE scenarios: Framework, storyline and drivers.

Patrick, D. A., Gibbs, J. P., Popescu, V. D., & Nelson, D. A. (2012). Multi-scale habitat-resistance models for predicting road mortality "hotspots" for turtles and amphibians. *Herpetological Conservation and Biology, 7*, 407–426.

Pereira, H. M., Leadley, P. W., Proença, V., et al. (2010). Scenarios for global biodiversity in the 21st century. *Science, 330*, 1496.

Pereira, H. M., Navarro, L. M., & Martins, I. S. (2012). Global biodiversity change: The bad, the good, and the unknown. *Annual Review of Environment and Resources, 37,* 25–50. doi:10.1146/annurev-environ-042911-093511.

Perkin, E. K., Hölker, F., Richardson, J. S., et al. (2011). The influence of artificial light on stream and riparian ecosystems: Questions, challenges, and perspectives. *Ecosphere, 2,* art122.

Philcox, C. K., Grogan, A. L., & Macdonald, D. W. (1999). Patterns of otter Lutra lutra road mortality in Britain. *Journal of Applied Ecology, 36,* 748–761.

Pressey, R. L., Humphries, C. J., Margules, C. R., et al. (1993). Beyond opportunism: Key principles for systematic reserve selection. *Trends in ecology & evolution, 8,* 124–128.

Proença, V., Pereira, H. M., & Vicente, L. (2010). Resistance to wildfire and early regeneration in natural broadleaved forest and pine plantation. *Acta Oecologica, 36,* 626–633.

Rentch, J. S., Fortney, R. H., Stephenson, S. L., et al. (2005). Vegetation–site relationships of roadside plant communities in West Virginia, USA. *Journal of Applied Ecology, 42,* 129–138.

Rey Benayas, J. M., Martins, A., Nicolau, J. M., & Schulz, J. J. (2007). Abandonment of agricultural land: An overview of drivers and consequences. *CAB reviews: Perspectives in agriculture, veterinary science, nutrition and natural resources, 2,* 1–14.

Rich, C., & Longcore, T. (2005). *Ecological consequences of artificial night lighting.* Washington, D.C.: Island Press

Ritchie, E. G., & Johnson, C. N. (2009). Predator interactions, mesopredator release and biodiversity conservation. *Ecology letters, 12,* 982–998.

Sanderson, E. W., Jaiteh, M., Levy, M. A., et al. (2002). The human footprint and the last of the wild. *BioScience, 52,* 891–904.

Schmitz, O. J. (2006). Predators have large effects on ecosystem properties by changing plant diversity, not plant biomass. *Ecology, 87,* 1432–1437.

Scottish, N. H. (2012). Wild land policy—Scottish Natural Heritage. http://www.snh.gov.uk/protecting-scotlands-nature/looking-after-landscapes/landscape-policy-and-guidance/wild-land/wild-land-policy/. Accessed 23 Aug 2013.

Seiler, A. (2005). Predicting locations of moose—Vehicle collisions in Sweden. *Journal of Applied Ecology, 42,* 371–382.

Sekercioglu, C. H. (2006). Increasing awareness of avian ecological function. *Trends in Ecology & Evolution, 21,* 464–471.

Selva, N., Kreft, S., Kati, V., et al. (2011). Roadless and low-traffic areas as conservation targets in Europe. *Environmental management, 48,* 865–877.

Sheffield, L. M., Crait, J. R., Edge, W. D., & Wang, G. (2001). Response of American kestrels and gray-tailed voles to vegetation height and supplemental perches. *Canadian Journal of Zoology, 79,* 380–385.

Sirami, C., Brotons, L., Burfield, I., et al. (2008). Is land abandonment having an impact on biodiversity? A meta-analytical approach to bird distribution changes in the north-western Mediterranean. *Biological Conservation, 141,* 450–459.

Sitch, S., Smith, B., Prentice, I. C., et al. (2003). Evaluation of ecosystem dynamics, plant geography and terrestrial carbon cycling in the LPJ dynamic global vegetation model. *Global Change Biology, 9,* 161–185.

Sydoriak, C. A., Allen, C. D., & Jacobs, B. F. (2000). Would ecological landscape restoration make the Bandelier Wilderness more or less of a wilderness. Proceedings: Wilderness science in a time of change conference, pp. 209–215.

Tabarelli, M., da Silva J. M. C., & Gascon, C. (2004). Forest fragmentation, synergisms and the impoverishment of neotropical forests. *Biodiversity & Conservation, 13,* 1419–1425.

Tacutu, R., Craig, T., Budovsky, A., et al. (2013). Human ageing genomic resources: Integrated databases and tools for the biology and genetics of ageing. *Nucleic acids research, 41,* D1027–D1033.

Valkama, J., Korpimäki, E., Arroyo, B., et al. (2005). Birds of prey as limiting factors of gamebird populations in Europe: A review. *Biological Reviews, 80,* 171–203.

Verburg, P. H., & Overmars, K. P. (2009). Combining top-down and bottom-up dynamics in land use modeling: Exploring the future of abandoned farmlands in Europe with the Dyna-CLUE model. *Landscape ecology, 24,* 1167–1181.

Vicente, J., Alves, P., Randin, C., et al. (2010). What drives invasibility? A multi-model inference test and spatial modelling of alien plant species richness patterns in northern Portugal. *Ecography, 33,* 1081–1092.

Vitousek, P. M., Aber, J. D., Goodale, C. L., & Aplet, G. H. (2000). Global change and wilderness science. Wilderness science in a time of change conference. RMRS-P-15-VOL-1, pp. 5–9.

Watson, J. E., Fuller, R. A., Watson, A. W., et al. (2009). Wilderness and future conservation priorities in Australia. *Diversity and Distributions, 15,* 1028–1036.

Wehn, S., Pedersen, B., & Hanssen, S. K. (2011). A comparison of influences of cattle, goat, sheep and reindeer on vegetation changes in mountain cultural landscapes in Norway. *Landscape and Urban Planning, 102,* 177–187.

Whittington, J., Hebblewhite, M., DeCesare, N. J., et al. (2011). Caribou encounters with wolves increase near roads and trails: A time-to-event approach. *Journal of Applied Ecology, 48,* 1535–1542.

Woolmer, G., Trombulak, S. C., Ray, J. C., et al. (2008). Rescaling the human footprint: A tool for conservation planning at an ecoregional scale. *Landscape and Urban Planning, 87,* 42–53.

Chapter 3
Ecosystem Services: The Opportunities of Rewilding in Europe

Yvonne Cerqueira, Laetitia M. Navarro, Joachim Maes,
Cristina Marta-Pedroso, João Pradinho Honrado and Henrique M. Pereira

Abstract Halting the degradation and restoring the full capacity of ecosystems to deliver ecosystem services is currently a major political commitment in Europe. Although still a debated topic, Europe's on-going farmland abandonment is seen as an opportunity to launch a new conservation and economic vision, through the restoration of natural processes via rewilding as a land management option. Despite the ecological interest of restoring a wilder Europe, there is a need to develop evidence-based arguments and explore the broad-range impacts of rewilding. In this chapter

Y. Cerqueira (✉) · L. M. Navarro · H. M. Pereira
Centro de Biologia Ambiental, Faculdade de Ciências da Universidade de Lisboa, Campo Grande, 1749-016 Lisboa, Portugal
e-mail: yvonnecerqueira@gmail.com

Y. Cerqueira · J. Pradinho Honrado
Centro de Investigação em Biodiversidade e Recursos Genéticos (CIBIO), Departamento de Biologia, Faculdade de Ciências da Universidade do Porto, 4169-007 Porto, Portugal

J. Pradinho Honrado
e-mail: joao.honrado@cibio.up.pt

L. M. Navarro · H. M. Pereira
German Centre for Integrative Biodiversity Research (iDiv) Halle-Jena-Leipzig, Deutscher Platz 5e, 04103 Leipzig, Germany

Institute of Biology, Martin Luther University Halle-Wittenberg, Am Kirchtor 1, 06108 Halle (Saale), Germany

L. M. Navarro
e-mail: laetitia.navarro@idiv.de

H. M. Pereira
e-mail: hpereira@idiv.de

J. Maes
European Commission, Joint Research Centre, Sustainability Assessment Unit, Via Fermi 2749, 21027 Ispra, VA, Italy
e-mail: joachim.maes@jrc.ec.europa.eu

C. Marta-Pedroso
IN+, Center for Innovation, Technology and Policy Research, Environment and Energy Scientific Area, Instituto Superior Técnico, 1049-001 Lisboa, Portugal
e-mail: cristina.marta@ist.utl.pt

H. M. Pereira, L. M. Navarro (eds.), *Rewilding European Landscapes,*
DOI 10.1007/978-3-319-12039-3_3, © The Author(s) 2015

we study the spatial patterns of ecosystem services in the EU25 and their relationship with wilderness areas. Next we perform a quantitative analysis, at the scale of the Iberian Peninsula, of the supply of ecosystem services in the top 5 % wilderness areas, on agricultural land, and on land projected to be abandoned. We find that high quality wilderness is often associated to high supply of ecosystem services, mainly regulating and cultural. Assuming that high quality wilderness is a good proxy for the future of areas undergoing rewilding, our results suggest that rewilding efforts throughout Europe will enhance the capacity of ecosystems to supply regulating and cultural ecosystem services, such as carbon sequestration and recreation.

Keywords Ecosystem services · Wilderness · Benefits · Farmland abandonment · Human well-being · Rewilding

3.1 Introduction

Ecosystem services have been defined as the benefits humans derive from nature through a set of ecosystem functions. The Millennium Ecosystem Assessment (MA 2005) was the stepping-stone in providing a conceptual framework for ecosystem services, which allows for assessing the consequences of ecosystem change for human well-being. Since its publication, multiple classification schemes for ecosystem services have been proposed, such as the framework of The Economics of Ecosystem Biodiversity (TEEB 2012) and, more recently, the Common International Classification of Ecosystem Services or CICES (Haines-Young and Potschin 2012). The CICES was adopted by the European Commission for the Mapping and Assessment of Ecosystem Services initiative (Maes et al. 2013). The CICES categorizes ecosystem services into 3 groups: provisioning (e.g. food, fiber, fuel and water), regulating and maintenance (e.g. air quality, water and soil regulation, natural hazard regulation, climate regulation and disease control) and cultural (e.g. recreation and spiritual).

Although society can easily perceive provisioning ecosystem services such as crops, fish and freshwater, which are all direct benefits to humans, others, such as pollination, erosion control and climate regulation are less tangible. However, directly or indirectly, all ecosystem services underpin environmental and human well-being, economy, and businesses (MA 2005). Many services are not traded in the conventional markets and hence, their economic values remain invisible, tending to be undervalued and consequently overexploited (de Groot et al. 2012). Yet, once lost, replacement can be costly. Wetlands, for example, provide numerous regulating services (e.g. water purification and flood/storm protection), which are unnoticed, in contrast to provisioning services (e.g. timber and food), but highly valuable since degradation can lead to high replacement costs (Reed et al. 2013).

Throughout the world, ecosystem services have been used as a tool in conservation and development as well as poverty alleviation (Tallis et al. 2008). The awareness that ecosystem services affect human well-being and economic development has resulted in their integration in the most recent EU Biodiversity Strategy (European Commission 2011a). This strategy aims at halting both biodiversity loss and the degradation of ecosystem services. It also includes the protection of wilderness, specifically old growth forest. Today, 45 % of Europe's land cover is forest (1 billion ha) but only 4 % is undisturbed forest (6 million ha). Protecting these ecosystems is important as they support particular ecosystem services, such as recreation and air quality (Maes et al. 2012a). Increasing the cover of wilderness areas in Europe trough rewilding of abandoned lands could improve the supply of these services (see Chap. 1). For instance, a recent initiative, "Rewilding Europe" aims at rewilding 1 million ha of land by 2020 (see Chap. 9). However, we have yet to determine what bundle of ecosystem services will rewilded areas provide.

In this chapter, we first investigate the supply and spatial distribution of ecosystem services on a pan-European scale. We then focus on the patterns of spatial overlap between the ecosystem services and wilderness areas. Next, we perform a quantitative analysis of the supply of services in the Iberian Peninsula, comparing the average supply between cultivated areas, high quality wilderness areas and areas currently cultivated but projected to be abandoned. Throughout the analysis, we consider the supply of ecosystem services in wilderness areas as a proxy for the future supply of services in rewilding areas. Finally, we discuss the various economic and ecological benefits of rewilding in Europe.

3.2   The Spatial Distribution of Ecosystem Services in Europe

Ecosystems provide a number of essential services underpinning all human life and activities. It is therefore important to recognize the multiple functions from ecosystems and integrate them in management strategies. To manage for multiple ecosystem services we need to map and identify the spatial synergies and trade-offs between services (Maes et al. 2012a; Raudsepp-Hearne et al. 2010). In doing so, we are able to identify ecosystems supporting high level of services and biodiversity (Chan et al. 2006). Along the years, the number of studies mapping ecosystem services has grown, informing both planners and decision makers on how to prioritize the protection and management of ecosystems (Chan et al. 2006; Naidoo and Ricketts 2006).

In the EU Biodiversity Strategy for 2020, the need for spatial assessment of ecosystem services has been included as one of the key actions. Under Action 5, all EU Member States are required to map and assess the state of ecosystems and their services by 2014. The results of this action will also contribute to the assessment of the economic value of ecosystem services, which is to be integrated into the accounting and reporting systems at both EU and national level by 2020 (European Commission 2011b).

Table 3.1 List of the ecosystem services and corresponding indicators used in the study. (Adapted from Maes et al. 2011). HANPP data were obtained from Haberl et al. (2007)

Service	Indicator	Unit	Description/benefit
Food provision	HANPP	$gC/m^2/year$	Human appropriation of net primary production (cropland and grassland in this study)
Timber provision	Total stock of timber	m^3/ha	Production for fuel, construction and paper. Forest connectivity
Freshwater provision	Surface water flow (QFS)	mm	Renewable freshwater provision
Climate regulation	Carbon stock	ton/ha	Above- and below-ground carbon stored in living plant material
	Net Ecosystem Productivity (NEP)	$mg/m^2/year$	Carbon sequestration
Water regulation	Nitrogen retention	%	Capacity of ecosystems to retain and process excess nitrogen
	Soil infiltration capacity	mm	Annual summed infiltration capacity of water
Air quality	Deposition velocity of Nox	cm/s	Capacity of ecosystems to capture and remove air pollutants
Recreation	Recreation potential index (RPI)	N/A	Capacity of ecosystems to provide recreational services

Here we build on on-going work to map ecosystem services across Europe (Maes et al. 2011). We consider a total of 7 ecosystem services, represented by 9 indicators (Table 3.1). In order for each ecosystem service to contribute equally to the analysis, and following the method of Petter et al. (2013), we standardized the data by reclassifying each service into a quantile split, producing a range of scores from 1 to 5 (five meaning high supply of a specific service). We then summed the 9 indicators to produce a map of "total" ecosystem services supply across Europe (Fig. 3.1a). We used the HANPP (Human Appropriation of Net Primary Production) data presented in Haberl et al. (2007), as the indicator for food provision. The HANPP values were only extracted within agricultural land as to not repeat the information on the provision of timber.

Low stocks for ecosystem service supply appear mainly around urbanized and densely populated areas and in arable land, e.g in central and eastern Spain, southern Romania, eastern UK, and Denmark (Fig. 3.1 a and b). However, low total supply of services does not mean a low quality of the supply of individual services. For example, even if food production were at their highest level in some areas, if that is the only service provided, such area would appear in the low range of the map. High total ecosystem service supply includes mainly pastures, forests and (semi) natural areas, such as the northwest Iberia, Scandinavia, central France, and central Romania. Areas of high total ecosystem service supply in Europe also coincide with mountain regions (Fig. 3.1a), mainly consisting of forest and (semi)-natural areas (Fig. 3.1b).

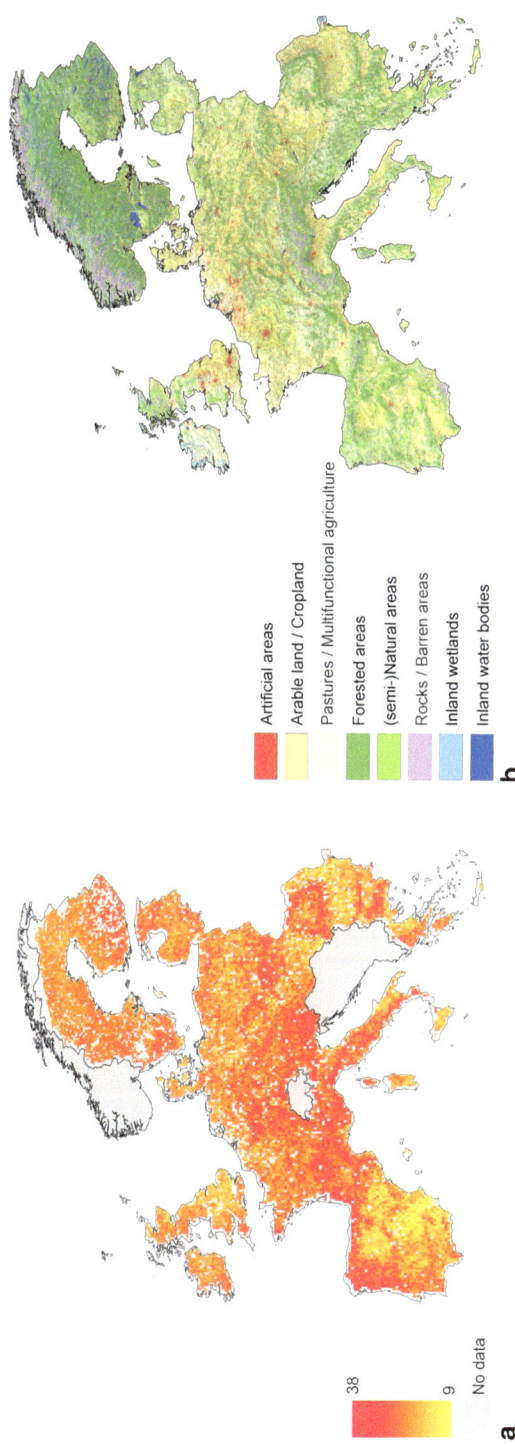

Fig. 3.1 Ecosystem services and land-covers in Europe. **a** Sum of the quantile splits of all indicators used in the analysis. By summing all 9 indicators (see Table 3.1), the gradient potentially varies between 9 and 45 but, de facto, the maximum and minimum values are 9 and 38 (Method detailed in the text). **b** Map of European land-covers based on the Corine Land Cover data base of 2006 (EEA 2010a). Both maps use an Eckert IV projection

Changes in human demand for services associated with specific land uses have shown diverging trends in Europe, varying between regions. In general the supply of crops, timber (mainly in northern countries), freshwater, and recreation has increased in the last 50 years while livestock production and wild foods supply have followed a decreasing trend throughout much of Europe's rural areas (Harrison et al. 2010). Other studies suggest that unsustainable farming practices and mismanagement through agricultural intensification have contributed to the loss of habitat and biodiversity, soil erosion and nutrient runoff (Dunbar et al. 2013).

3.3 Wilderness, Rewilding and Ecosystem Services

Wilderness

Wilderness areas have been defined as large natural areas, unmodified or slightly modified, governed by natural processes, with no human intervention, infrastructure or permanent habitation present (Wild Europe 2012). Nordic mountains represent the highest proportion (28 %) of wildest areas, followed by the Pyrenees (12 %) the eastern Mediterranean islands and Alps (9 %), and British Isles (8 %) (Carver 2010). However, remnants can also be found throughout much of the continent, where anthropogenic interference has slightly altered the natural ecological conditions (Carver 2010). The definition of wilderness will depend on the metrics chosen (see Chap. 2) and, as a result, its spatial distribution can vary from one study to another. Currently, there are several maps on potential wilderness in Europe. Here we chose to use Carver's (2010) quality wilderness index.

Wild ecosystems provide a wide range of ecosystem services. They are stable and self-sustainable, able to maintain their structure, function and resilience over time (Costanza and Mageau 1999). They play an important role in protecting services such as, air quality, freshwater provision, and supporting wildlife, including charismatic species, such as bisons and bears, that are reliant on wilderness areas (Russo 2006; see Chap. 9). Wild ecosystems also have the capacity to supply higher quality services than other types of systems. For example, there is higher carbon storage capacity in undisturbed forest, peatland and wetland (Schils et al. 2008), subsequently providing additional environmental benefits (e.g. biodiversity, water storage and water quality). Moreover, wilderness areas provide a range of social and economic benefits. Several programs have integrated the use of wild areas to address urban issues such as youth at risk, youth development and rehabilitation (Hill 2007), and recognized it as a cost-effective form of healthcare. In addition, wilderness inspires educational programs (e.g. Chap. 10). Wilderness areas also provide spiritual benefits, such as, solitude, places of inspiration, a calm environment, and recreation/tourism (Ewert et al. 2011; Heintzman 2013). These cultural services can give birth to employment opportunities and thus generate income. For example, the Oulanka National Park in Finland brings 14 million € per year to the local economy and employs 183 individuals (Huhtala et al. 2010).

Methods

We used Carver's (2010) wilderness quality map and the same quantile approach described earlier in Sect. 3.2, to produce a gradient of wilderness quality with qualitative values ranging between 1 and 4 (4 meaning the highest wilderness quantile and high supply of ecosystem services). We then grouped the ecosystem services into provisioning, regulating, and cultural services and followed the same splitting approach for each group of services. The ecosystem services maps were then overlaid with the wilderness map. To determine the relationship between gradients of both ecosystem services supply and wilderness quality, we display the overlay of high and low wilderness with high and low supply of ecosystem services (Fig. 3.2a, b, c and d). Furthermore, we used the projections of the CLUE model (Verburg and Overmars 2009) to assess the potential change in the provision of ecosystem services with scenarios of land abandonment and rewilding in Europe for 2030. We considered as potential land abandonment and rewilding the cells classified as arable land, pasture, irrigated arable land, permanent crops in 2000 and classified in 2030 in all four EURURALIS scenarios as (semi)-natural vegetation, forest, recently abandoned arable land and recently abandoned pasture land. For quantitative comparisons, we calculated the mean provision of ecosystem service (per km^2) in agricultural areas (based on the 2000 land use map, in Verburg and Overmars 2009), in the top 5% high quality wilderness, and in the areas currently under agricultural use but projected to become abandoned by 2030, in the Iberian Peninsula (Table 3.2). Differences between the distributions of the mean ecosystem service values for each type of land-use were assessed using a Kruskal-Wallis test. Finally, we calculated the ratio between the average supply of each indicator in either the top 5% wilderness areas and in agricultural areas relative to the areas projected to be abandoned. All mapping and data extraction were done using ArcGIS version 10.3, while the statistical analysis was done using R version 2.15.3.

Wilderness and Ecosystem Services

Some high wilderness areas are associated to regions supplying high ecosystem services, particularly in mountain regions (Fig. 3.2a). As expected, the overlay of provisioning services and wilderness (Fig. 3.2b) exhibits relatively large areas of high supply of services and low wilderness (e.g. in France, Benelux and Germany), along with areas of low service supply and high wilderness (e.g. Northern Scandinavia). This is not surprising since wilderness areas are typically associated with low to no extraction of natural resources. There are nonetheless high provisioning services in some areas of high wilderness quality, mainly associated to mountain regions (e.g. some areas of the Alps and Apennines). This can be due to the occurrence of large quantities of resources for some provisioning services (i.e. timber and freshwater) in mountain regions, which still happen to be wilder than the rest of Europe.

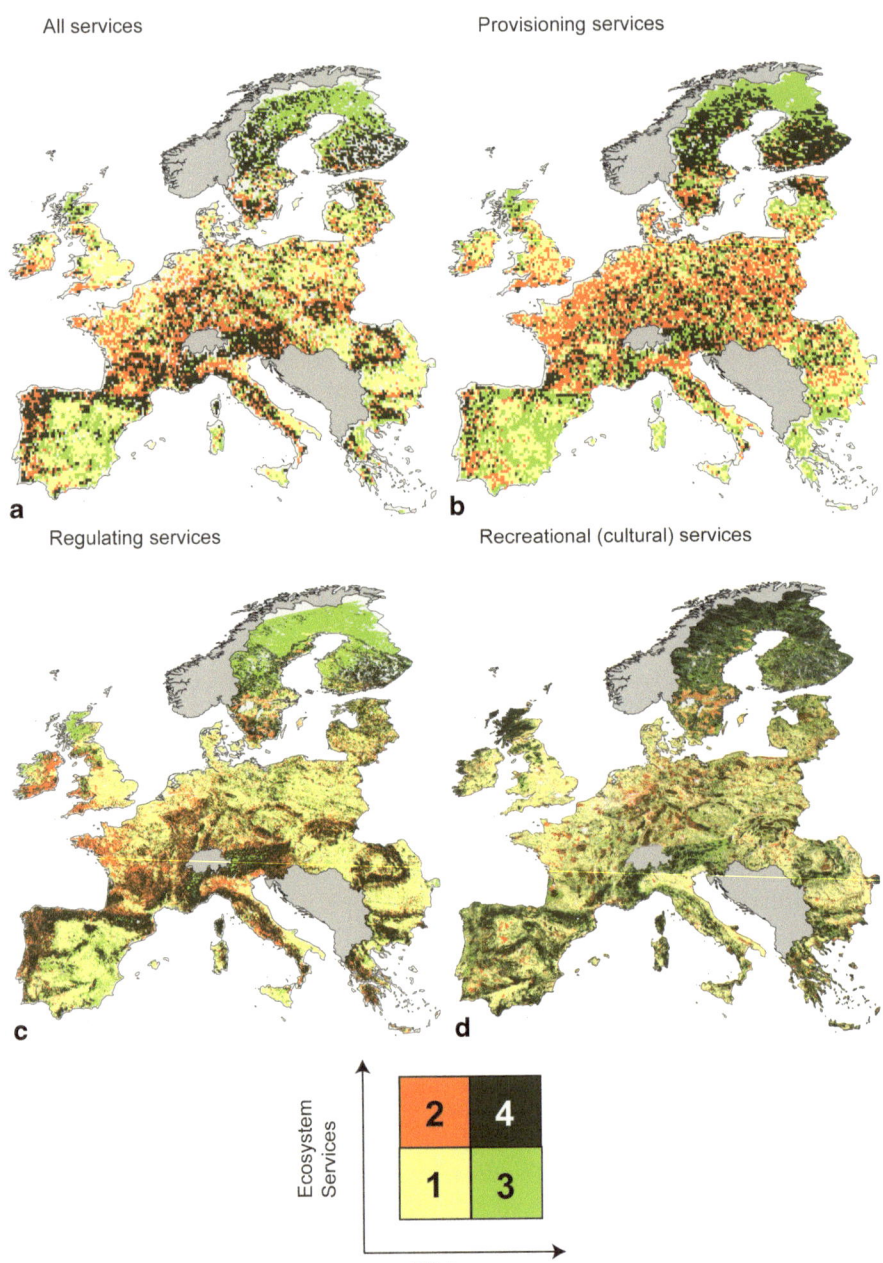

Fig. 3.2 Ecosystem services and wilderness in Europe. For each map, the quantile splits of ecosystem services and wilderness were overlaid to present a gradient of both wilderness and service supply. For an easier representation, the values were grouped into "low" (*bottom* 50%) and "high" (*top* 50%) for both metrics and then grouped, e.g. low supply of services and low wilderness (see color key on the figure). **a** All indicators for all services versus wilderness; **b** Indicators of provisioning services versus wilderness; **c** Indicators of regulating services versus wilderness; and **d** Recreational service versus wilderness. (See Table 3.1 for details on the indicators used). (Sources: Carver 2010; Maes et al. 2011)

Table 3.2 Quantitative analysis of the supply of ecosystem services in the Iberian Peninsula, on agricultural land, top 5 % high wilderness areas (Carver 2010) and land currently cultivated and projected to be abandoned by 2030 (See Mapping methods section). The p value of the Kruskal Wallis test and the level of significance are given for each indicator: ***, $p < 0.001$ and NS, $p > 0.05$

Category	Service	Indicator	Unit	Mean per km^2 (sem)			p value
				Agricultural area	Top 5 % high wilderness	Projected to be abandoned	
Provision-ning	Food production	HANPP	gC/m^2/yr	**279.03**(0.21)	224.12(2.65)	271.15(2.91)	<0.0001(***)
	Timber	total stock	m^3/ha	0.81E+05 (0.21E+03)	**2.10 E+05** (1.67E+03)	1.34E+05 (1.68E+03)	0.23 (NS)
	Freshwater	Surface water flow	mm	151.05 (0.15)	156.08 (0.58)	**244.77** (1.61)	<0.0001(***)
Regula-tion and Maintenance	Climate regulation	Carbon stock	ton/ha	24.82 (0.06)	63.19 (0.30)	**74.81** (0.49)	<0.0001(***)
		Net Ecosystem Productivity (NEP)	mg/m^2/yr	5.13E+05 (0.34E+03)	6.99 E+05 (1.75E+03)	**7.36E+05** (1.92E+03)	<0.0001(***)
	Water regulation	Nitrogen retention	%	2.69 (.00)	**3.04** (.01)	2.46 (.02)	<0.0001(***)
		Soil infiltration capacity	mm	9.61 (0.02)	15.99 (0.11)	**32.15** (0.23)	0.0006(***)
	Air quality	Deposition veloc-ity of Nox	cm/s	0.07 (0.00)	**0.42** (0.00)	0.28 (0.01)	<0.0001(***)
Cultural	Recreation	Recreational Potential Index (RPI)	N/A	0.22 (2E-04)	**0.43** (11E-04)	0.26 (17E-04)	<.0001(***)

The spatial distribution of wilderness coincides more with regulating services than with provisioning services (Fig. 3.2c). Areas of Europe containing both high supply of services and high degrees of wilderness include mountainous areas in Northern Iberia, Austria, and Italy. Most of the continent is still represented by areas of both low regulating services and low wilderness (e.g. Eastern UK, Poland), which also coincides with agricultural areas (Fig. 3.1b). Interestingly, several areas of high supply of regulating services and low wilderness also exist (Western France and Ireland).

Finally, for recreational services (Fig. 3.2d), we found a predominance of either areas of low service supply and low wilderness, or areas of high wilderness and high service supply, suggesting this is the category of services most strongly associated with wilderness. However there are some areas of low wilderness and high service supply or areas with high wilderness but low recreation potential. Typically, the flow of recreational services is calculated as the product between the capacity of an area to supply recreational services and the accessibility of this area (Maes et al. 2011). As a result, it can occur that an ecosystem would be of extreme beauty or wilderness quality but not accessible, leading to a low flow of recreation and other cultural services, or that an area would be less natural but still be an important cultural landscape that is easily accessible.

Taken as a whole, regulating and cultural services are often associated to high wilderness areas (Fig. 3.2b and c), particularly mountain systems. Mountain ecosystems cover approximately 41 % of Europe's territory, providing various services due to their multifunctionality. Mountains are "water towers" as they provide water for multiple uses, including irrigation, human consumption, and hydropower (Viviroli et al. 2007). Mountain systems supply cultural services, holding spiritual value to local inhabitants, and are recreation and ecotourism attractions (Price et al. 1997). In mountain systems there is a high proportion of habitat types with favorable conservation status (EEA 2010b), playing a key role in supplying many ecosystem services and maintaining ecological processes (Harrison et al. 2010). Forests make up 41 % of mountain systems (Körner et al. 2005) and can be regulators of natural disasters as the soils of mature forests have high infiltration rate, thus reducing peak flows and floods (Maes et al. 2009). Forests also provide a range of services such as carbon sequestration, air quality regulation, timber for fuelwood and non-timber products (game and medicinal plants), and climate regulation (Harrison et al. 2010; Maes et al. 2012b). Peatlands store large quantities of carbon and have played a fundamental role in climate regulation and are critical for water regulation. Grasslands are the habitat of a large number of species, such as wild pollinators (Kremen et al. 2002), which makes them essential in underpinning biodiversity and ecosystem services. Finally, mountains are also hotspots of endemism. In Europe, the highest number of endemic species can be found in the Alps and the Pyrenees (Väre et al. 2003).

Ecosystem Services and Scenarios of Rewilding

Here we estimate the biophysical potential of rewilding to produce benefits, by comparing ecosystem services in the top 5% wilderness areas with the current supply of ecosystem services in all agricultural areas and in agricultural areas that are projected to be abandoned. We restrict the analysis to the scale of the Iberian Peninsula to control for the large bioclimatic variability across Europe.

There are significant differences in the supply of most ecosystem services between the different land use categories (Table 3.2). HANPP values are significantly higher in agricultural areas than in both land projected to be abandoned and, as expected, in the top 5% wilderness areas. We thus hypothesize that food production will decrease with the contraction of the agricultural area, although the decrease will be limited because of the lower agricultural productivity of those areas. Several services present higher values for the average supply of the studied indicators in the top 5% wilderness areas (Table 3.2). The deposition velocity of NOx, an indicator of air quality, depends on the height of the vegetation and the leaf area index, and tends to be much higher in forested areas (Maes et al. 2011), hence, the higher values in the top 5% wilderness (Table 3.2). The recreation potential is also higher in wilderness areas than in the other land-uses.

Most ecosystem services exhibit higher values in the areas to be abandoned than in other agricultural areas (Fig. 3.3a). We can thus speculate that intensifying agriculture in the areas projected to be abandoned would lead to an overall decrease in the supply of ecosystem services in these areas. On the other hand, rewilding these areas would bring improvements in some ecosystem services, such as nitrogen retention and recreation, and decreases in others (Fig. 3.3b). These inferences have to be interpreted with care as we are making several simplifying assumptions and ecosystem services depend on other biophysical variables besides land cover and

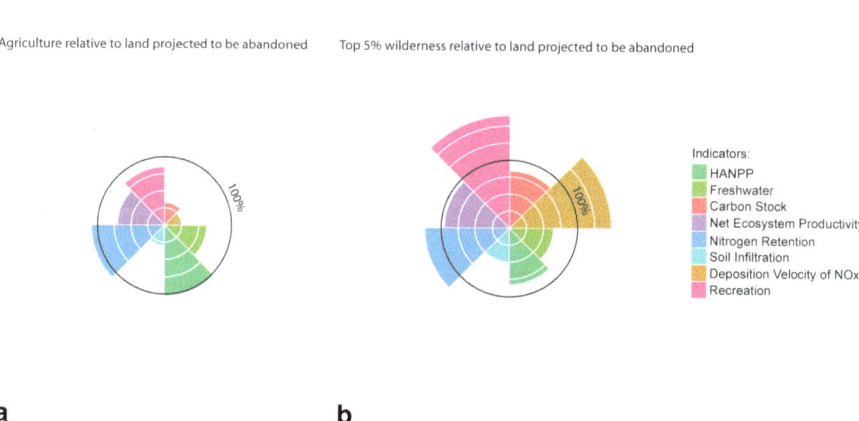

Agriculture relative to land projected to be abandoned   Top 5% wilderness relative to land projected to be abandoned

Indicators:
HANPP
Freshwater
Carbon Stock
Net Ecosystem Productivity
Nitrogen Retention
Soil Infiltration
Deposition Velocity of NOx
Recreation

a   **b**

Fig. 3.3 Comparison between land-uses of the average supply of indicators of ecosystem services. The diagrams represent the average supply per km² in cultivated areas (**a**), or in the top 5% wilderness (**b**), relative to the average supply in areas projected to be abandoned. Values inside the 100% circle are lower for the studied land-use, while values outside are higher when compared with land projected to be abandoned. See Table 3.2 for the average values.

land use. This is for instance the case for freshwater supply, which depends on the aboveground net precipitation water in catchments and on the area and flows of the freshwater areas (Maes et al. 2011).

3.4   The Economic Benefits of Rewilding

In the previous sections we used models of the spatial distribution of ecosystem services to look at the potential impacts of rewilding on biophysical metrics of indicator supply. We now review case studies of economic valuation of ecosystems services provided by natural habitats and by ecological restoration.

Regulating Benefits

In Lowland England, studies on different land use management options have shown that the cost and benefits of changes in ecosystem services from rewilding outweigh those from arable and dairy farming (NERC 2012). In the Upland UK estimates show that managing the land for carbon storage and sequestration through the restoration of peatlands may be more profitable than pastoral activities (Reed et al. 2013). Peatlands, in Scotland, have been valued between 49 million € and 196 € per annum for carbon sequestration (McMorran et al. 2006).

Forest regeneration will also provide major increases in carbon sequestration. It has been estimated that within the Natura 2000 network, commercial and wild forest habitats generate the highest carbon value estimated at 318.3 € and 610.1 billion €, in 2010 followed by grassland systems ranging between 105.6 € and 196.5 billion € (ten Brink et al. 2011). In the Carpathians, the protection of old growth forest is expected to generate 26 million € through carbon offsets (ten Brink et al. 2011). In the Hoge Veluwe Forest, a protected area of the Netherlands, total economic benefit generated by forests is 2000 € ha/year, for the following services: wood production, supply of game, groundwater recharge, carbon sequestration, air filtration, recreation and nature conservation. This value is calculated to be three times higher than adjacent agricultural land (Hein 2011).

Although there is still a lack of available information on the economic value of water purification at the EU level, studies suggest that cities such as Berlin, Vienna, Oslo, and Munich benefit from the natural treatment from ecosystems in protected and non protected areas, with annual economic benefits ranging between 7 € and 16 million € for water purification and 12 € and 91 million € for water provision per city (ten Brink et al. 2011). In the archipelago of the Azores, the restoration of pastures to native forests would result in an economic benefit of 110 € thousand per year from water purification (Cruz and Benedicto 2009). These examples, though limited, demonstrate that protecting and restoring natural vegetation is of economic benefit, and could contribute to achieve the goals of the Water Framework Directive.

Floodplains (wetlands) are also important ecosystems for water cycle regulation, acting as natural sponges, they retain water in river basins, slowly releasing the water down river and into groundwater. Moreover, they play a fundamental role in filtering out pollutants and are home to much wildlife. Restoring the function of floodplains in EU countries could save approximately 1.4 billion € of treatment costs for water purification and reduce annual cost of flood damage, currently at 6.4 billion € and expected to increase (Feyen and Watkiss 2011). Of course, this type of ecosystem restoration has initial costs. The Danube Basin restoration project estimates that the recovery of 100,000 ha, would cost 500,000 €/km², i.e. an investment of 500 million €. However, this value is still estimated to be much lower than the costs associated to damage control and the improvement of dykes (WWF 2010).

Degradation of natural ecosystems has also been linked to the intensification of other natural hazards (Dudley et al. 2010). For example, in the Swiss Alps the protection of old forests contribute to disaster prevention (e.g. avalanches and landslides) and have been analyzed at a value of 1.6–2.8 billion € per year (IPCDR 2010). Additionally, the role of European pristine scrublands and Belgian grasslands against soil erosion was valued at 44.5 €/ha (Kettunen et al. 2012).

Cultural Benefits

Economic benefits from non-extractive activities such as nature tourism and recreation boost local and regional economies, providing income and employment to communities and private landholders who face limited alternative livelihoods, especially in a context of rural depopulation of marginal areas (Brown et al. 2011; McMorran et al. 2006). Furthermore, the aims of eco-tourism are closely associated with biodiversity conservation. Through the promotion of rewilding efforts, there will be an increase in the connectivity of landscapes, creating an opportunity for the expansion of large mammals and other species (Russo 2006), and indirectly increasing tourism while generating economic benefits to local communities.

Presently, eco-tourism is the fastest growing component sector in tourism (Gössling 2000). Overall, tourism is the largest global economic sector accounting for $ 3.6 trillion in economic activity and eco-tourism has constantly increased 20–30 % per year since the early 1990's (Bishop et al. 2008). Eco-tourism is defined by the International Ecotourism Society as the responsible visiting to natural areas that conserves the environment and improves the well-being of local people. For instance, in Zarnesti, Romania, a small community increased their total local revenue from 140,000 € in 2001 to 260,000 € in 2002 through eco-tourism programmes (CLCP 2000).

In particular, wildlife areas appeal to a large spectrum of tourists given the presence of charismatic species and other rare or attractive species. For example the reintroduction of wolves in the Yellowstone National Park has attracted additional tourists, generating economic and social benefits estimated at US$ 6–9 million per year (Donlan et al. 2006). The reintroduction of ungulates and large carnivores in the Majella and the Retezat National Park in Italy and Romania, respectively, has also contributed to the local economy (Kun and van der Donk 2006). In Scotland, tourism from wild landscapes is one of the most important economic sectors, contributing 1.6 billion € annually, to the country's economy. In particular, recreation opportunities, such as wildlife watching and hillwalking, generate 65 million € and support 39,000 full time jobs (Brown et al. 2011; Bryden et al. 2010). The reintroduction of the beaver can potentially generate an additional £ 2 million per year into the local Scottish economy through eco-tourism (Campbell et al. 2007). In addition to its potential economic benefits, beaver dams are considered to have a positive impact on river systems by increasing both invertebrate and fish populations (Kemp et al. 2010).

The Natura 2000 network further exemplifies how biodiversity can be protected while generating benefits. Annually, the gross socio-economic and co-benefits (social and environmental) from the Natura 2000 network range between 223 billion € and 314 billion €, representing between 2 and 3 % of EU's GDP (ten Brink et al. 2011). This figure contrasts with the annual investment in the Natura 2000 network, estimated at 5.8 billion € while providing 8 million (FTE) jobs (Gantolier et al. 2010).

3.5 Discussion

The degradation, or land conversion, of natural ecosystems alters not only species richness and composition; it reduces ecosystem functionality, impacting the flow of ecosystem services, the costs of recuperation and ultimately human well-being (Flynn et al. 2009). Global and EU targets were designed for the conservation and restoration of natural ecosystems, including the biodiversity and ecosystem services that they sustain (see Chap. 11). For instance, Target 2 of the EU 2020 biodiversity strategy promotes the restoration and the use of green infrastructures (i.e interconnected network of ecosystems, such as wetlands and woodlands) with the goal of restoring 15 % of degraded ecosystems, through incentives based on EU funding and Public Private Partnerships (European Commission 2011a). In this context, the restoration of nature through rewilding can be seen as a solution to address the on-going agricultural land abandonment while developing a new rural economy offering multiple social and environmental benefits (Brown et al. 2011; Bryden et al. 2010; Donlan et al. 2006; Gantolier et al. 2010; Hein 2011; McMorran et al. 2006).

We investigated the existence of the spatial co-occurrence of wilderness and ecosystem services supply at the EU scale (Fig. 3.2). Our results further suggest that the opportunity of restoring abandoned land in the Iberian Peninsula to a self-sustained natural state, via rewilding, could increase the supply of regulating and cultural services (Table 3.2 and Fig. 3.3). We thus argue that by restoring and sustaining wilderness areas we are underpinning a supply of high quality ecosystem services provided by those areas. These services will also heighten a new local economy, providing an economic break for the remaining rural communities through the creation of jobs and income generated from incentives, including from payments for ecosystem services, carbon markets, biodiversity markets, and eco-tourism (e.g.

Bishop et al. 2008; Jack et al. 2008; Pirard 2012; TEEB 2010). Although, the concept of rewilding is fairly recent in Europe, it has already been identified as a cost-effective management strategy for traditional land uses in Scotland (Brown et al. 2011; McMorran et al. 2006). In the Netherlands, rewilding has been positively perceived by people: individuals attribute a low willingness to pay for the conservation of extensive farming versus rewilding initiatives (van Berkel and Verburg 2014).

Farmland abandonment can lead to the potential loss of traditional cultural values and heritage, including local knowledge on farming and resource management, and locally adapted animal breeds and crop varieties (Cerqueira et al. 2010). Thus, choices have to be made case by case, and strategies should be designed to mitigate and avoid cultural losses. Furthermore, extensive agriculture and the maintenance of traditional activities provide a different bundle of ecosystem services from rewilding. Therefore, there might be instances where local communities or the public will prefer the bundle of services associated with rewilding, while in other places the bundle of services associated with extensive agriculture will be chosen.

In conclusion, we are not suggesting that rewilding efforts through assisted or passive restoration be the only solution to Europe's present situation. Instead we think it should be considered as a potential strategy in those areas where the social-ecological dynamics of the landscape are no longer socially, economically or environmentally sustainable. Yet, there are still many challenges in understanding the full relationship between landscape management, the supply of ecosystem services, and the economic benefits and costs associated to each management type. We believe we need further research on the environmental, social and economic benefits associated to wilderness and rewilded areas. Raising awareness of these benefits may help to promote the concept of rewilding, and help gain momentum to define public policies and funding for rewilding activities.

Acknowledgments We thank Vânia Proença and Alexandra Marques for insightful comments on earlier versions of the manuscript. L.M.N. was supported by a PhD fellowship from the FCT (SFRH/BD/62547/2009).

References

Bishop, J., Kapila, S., Hicks, F., Mitchell, P., & Vorhies, F. (2008). *Building biodiversity business* (164 pp). London: Shell International Limited and the International Union for Conservation of Nature.

Brown, C., Mcmorran, R., & Price, M. F. (2011). Rewilding–A new paradigm for nature conservation in Scotland? *Scottish Geographical Journal, 127*(4), 288–314.

Bryden, D. M., Westbrook, S. R., Burns, B., Taylor, W. A., & Anderson, S. (2010). *Assessing the economic impacts of nature based tourism in Scotland—Commissioned Report No. 398*. Scottish Natural Heritage.

Campbell, R. D., Dutton, A., & Hughes, J. (2007). *Economic impacts of the beaver*. Report for the Wild Britain Initiative. Oxon, UK, 28 p.

Carver, S. (2010). Chapter 10.3 Mountains and wilderness. In Europe's ecological backbone: Recognising the true value of our mountains (No. 6/2010). European Environment Agency. Copenhagen, Denmark, 248 pp.

Cerqueira, Y., Araujo, C., Vicente, J., Pereira, H. M., & Honrado, J. (2010). Ecological and cultural consequences od agricultural abandonment in the Peneda-Gerês National Park (portugal). In N. Evelpiou, et al. (Eds.), *Natural Heritage from East to West*. (pp. 175–183). Berlin: Springer.

Chan, K., Shaw, M., Cameron, D., Underwood, E., & Daily, G. (2006). Conservation planning for ecosystem services. *PLoS Biology, 4*(11), 2138–2152. doi:10.1371/journal.pbio.0040379.

CLCP. (2000). *Annual Report*. Zarnesti Romania: Carpathian Large Carnivore Project.

Costanza, R., & Mageau, M. (1999). What is a healthy ecosystem? *Aquatic Ecology, 33*(1), 105–115. doi:10.1023/A:1009930313242.

Cruz, A. de la, & Benedicto, J. (2009). Assessing Socio-economic Benefits of Natura 2000—A Case Study on the ecosystem service provided by SPA PICO DA VARA/RIBEIRA DO GUILHERME. Output of the project Financing Natura 2000: Cost estimate and benefits of Natura 2000 (Contract No.: 070307/2007/484403/MAR/B2). 43 pp.

De Groot, R., Brander, L., van der Ploeg, S., Costanza, R., Bernard, F., Braat, L., & van Beukering, P. (2012). Global estimates of the value of ecosystems and their services in monetary units. *Ecosystem Services, 1*(1), 50–61. doi:10.1016/j.ecoser.2012.07.005.

Donlan, C. J., Berger, J., Bock, C. E., Bock, J. H., Burney, D. A., Estes, J. A., et al. (2006). Pleistocene rewilding: An optimistic agenda for twenty-first century conservation. *American Naturalist, 168*(5), 660–681.

Dudley, N., Stolton, S., Belokurov, A., Krueger, L., Lopoukhine, N., MacKinnon, K., Sandwith, T., & Sekhran N. (Eds.). (2010). *Natural solutions: Protected areas helping people cope with climate change*. Gland: IUCN, WCPA, TNC, UNDP, WCS, The World Bank and WWF.

Dunbar, M. B., Panagos, P., & Montanarella, L. (2013). European perspective of ecosystem services and related policies. *Integrated Environmental Assessment and Management, 9*(2), 231–236.

EEA. (2010a). *Corine Land Cover 2006 raster data*. European Environment Agency.

EEA. (2010b). *Integrated assessment of Europe's mountain areas*. Copenhagen: European Environment Agency.

European Commission. (2011a). *Our life insurance, our natural capital: An EU biodiversity strategy to 2020 (No. COM (2011) 244 final)*. Brussels: European Comission.

European Commission. (2011b). *The EU Biodiversity Strategy to 2020*. Luxembourg, 27pp,

Ewert, A., Overholt, J., Alison, V., & Wang, C. (2011). *Understanding the Transformative Aspects of the Wilderness and Protected Lands Experience upon Human Health*. USDA Forest Service Proceedings RMRS-P-64: P. 140-146.

Feyen, L., & Watkiss, P. (2011). *Technical policy briefing note 3: The impacts and economic costs of river floods in Europe, and the costs and benefits of adaptation*. Sweden: Stockholm Environment Institute.

Flynn, D. F., Gogol-Prokurat, M., Nogeire, T., Molinari, N., Richers, B. T., Lin, B. B., & DeClerck, F. (2009). Loss of functional diversity under land use intensification across multiple taxa. *Ecology Letters, 12*(1), 22–33.

Gantolier, S., Rayment, M., Bassi, S., Kettunen, M., McConville, A., Landgrebe, R., & ten Brink, P. (2010). *Costs and socio-economic benefits associated with the natura 2000 network. Final report to the European commission. (No. G Environment on Contract ENV.B.2/SER/2008/0038.)* (p. 181). Brussels: Institute for European Environmental Policy/GHK/Ecologic.

Gössling, S. (2000). Tourism—sustainable development option? *Environmental Conservation, 27*(03), 223–224.

Haberl, H., Erb, K. H., Krausmann, F., Gaube, V., Bondeau, A., Plutzar, C., & Fischer-Kowalski, M. (2007). Quantifying and mapping the human appropriation of net primary production in earth's terrestrial ecosystems. *Proceedings of the National Academy of Sciences, 104*(31), 12942–12947.

Haines-Young, R., & Potschin, M. (2012). *CICES V4.3—Report prepared following consultation on CICES Version 4*. EA Framework Contract No EEA/IEA/09/003.

Harrison, P. A., Vandewalle, M., Sykes, M. T., Berry, P. M., Bugter, R., de Bello, F., & Haslett, J. R. (2010). Identifying and prioritising services in European terrestrial and freshwater ecosystems. *Biodiversity and Conservation, 19*(10), 2791–2821.

Hein, L. (2011). Economic benefits generated by protected areas: The case of the Hoge Veluwe forest, the Netherlands. *Ecology & Society, 16*(2), 13.

Heintzman, P. (2013). Spiritual outcomes of park experience: A synthesis of recent social science research. *George Wright Forum, 30,* 273–279.

Hill, N. R. (2007). Wilderness therapy as a treatment modality for at-risk youth: A primer for mental health counselors. *Journal of Mental Health Counseling, 29*(4), 338–349.

Huhtala, M., Kajala, L., & Vatanen, E. (2010). *Local economic impacts of national park visitors' spending in Finland: The development process of an estimation method.* Working papers of the Finnish forest research institute 149. http://www.researchgate.net/publication/228455705_Local_economic_impacts_of_national_park_visitors'_spending_in_Finland_The_development_process_of_an_estimation_method/file/50463528343f6c63cc.pdf.

IPCDR. (2010). Promoting Payments for Ecosystem Services in the Danube Basin. *Danube Watch-the Magazine of the Danube River, 3.*

Jack, B. K., Kousky, C., & Sims, K. R. E. (2008). Designing payments for ecosystem services: Lessons from previous experience with incentive-based mechanisms. *Proceedings of the National Academy of Sciences, 105*(28), 9465–9470. doi:10.1073/pnas.0705503104.

Kemp, P. S., Worthington, T. A., & Langford, T. E. L. (2010). A critical review of the effects of beavers upon fish and fish stocks. Scottish Natural Heritage Commissioned Report No. 349 (iBids No. 8770).

Kettunen, M., Vihervaara, P., Kinnunen, S., D'Amato, D., Badura, T., Argimon, M., & ten Brink, P. (2012). *Socio-economic importance of ecosystem services in the Nordic Countries.* The Economics of Ecosystems and Biodiversity (TEEB).

Körner, C., Spehn, E., & Baron, J. (2005). Mountain systems. In W. R. Institute (Ed.), *Millennium ecosystem assessment: Ecosystems and human well-being: Synthesis.* Island Press, Washington DC.

Kremen, C., Williams, N. M., & Thorp, R. W. (2002). Crop pollination from native bees at risk from agricultural intensification. *Proceedings of the National Academy of Sciences, 99*(26), 16812–16816.

Kun, Z., & van der Donk, M. (2006). Providing wilderness experience opportunities in Europe's certified PAN parks. *Parks, 16*(2), 34–40.

MA. (2005). *Millennium ecosystem assessment: Ecosystems and human well-being: Synthesis.* Washington, D.C.: Island Press.

Maes, W. H., Heuvelmans, G., & Muys, B. (2009). Assessment of land use impact on water-related ecosystem services capturing the integrated terrestrial- aquatic system. *Environmental Science & Technology, 43*(19), 7324–7330.

Maes, J., Paracchini, M. L., & Zulian, G. (2011). *A European assessment of the provision of ecosystem services: Towards an Atlas of ecosystem services.* Luxembourg: European Union.

Maes, J., Paracchini, M. L., Zulian, G., Dunbar, M. B., & Alkemade, R. (2012a). Synergies and trade-offs between ecosystem service supply, biodiversity, and habitat conservation status in Europe. *Biological Conservation, 155,* 1–12.

Maes, J., Hauck, J., Paracchini, M. L., Ratamäki, O., Termansen, M., Perez-Soba, M., et al. (2012b). *A spatial assessment of ecosystem services in Europe: Methods, case studies and policy analysis-phase 2 Synthesis report* (PEER Report No. 4).

Maes, J., Teller, A., Erhard, M., Liquete, C., Braat, L., Berry, P., & Bidoglio, G. (2013). *Mapping and assessment of ecosystems and their services. An analytical framework for ecosystem assessments under action 5 of the EU biodiversity strategy to 2020.* Luxembourg: European Comissions.

McMorran, R., Price, M. F., & McVittie, A. (2006). A review of the benefits and opportunities attributed to Scotland's landscapes of wild character. Scottish Natural Heritage Commissioned Report No. 194 (ROAME No. F04NC18).

Naidoo, R., & Ricketts, T. H. (2006). Mapping the economic costs and benefits of conservation. *PLoS Biology, 4*(11), e360.

NERC. (2012). *Valuing ecosystem services: Case studies from lowland England. Annex 4- Knepp Castle Estate Re-wilding:Sussex*. Natural England.

Petter, M., Mooney, S., Maynard, S. M., Davidson, A., Cox, M., & Horosak, I. (2013). A Methodology to map ecosystem functions to support ecosystem services assessments. *Ecology & Society, 18*(1), 31.

Pirard, R. (2012). Market-based instruments for biodiversity and ecosystem services: A lexicon. *Environmental Science & Policy, 19–20*, 59–68. doi:10.1016/j.envsci.2012.02.001.

Price, M. F., Moss, L. A., & Williams, P. W. (1997). Tourism and amenity migration. In B. Messerli & J. D. Ives (Eds.), *Mountains of the world: A global priority* (pp. 249–280). Parthenon Publishing Group, New York.

Raudsepp-Hearne, C., Peterson, G. D., & Bennett, E. M. (2010). Ecosystem service bundles for analyzing tradeoffs in diverse landscapes. *Proceedings of the National Academy of Sciences, 107*(11), 5242–5247. doi:10.1073/pnas.0907284107.

Reed, M. S., Hubacek, K., Bonn, A., Burt, T. P., Holden, J., Stringer, L. C., et al. (2013). Anticipating and managing future trade-offs and complementarities between ecosystem services. *Ecology and Society, 18*(1), 5.

Russo, D. (2006). Effects of land abandonment on animal species in Europe: Conservation and management implications Integrated assessment of vulnerable ecosystems under global change in the EU. Project report. 52 pp. *Università Degli Studi de Napoli Federico, Napoli, Italy*.

Schils, R., Kuikman, P., Liski, J., Van Oijen, M., Smith, P., Webb, J., & Hiederer, R. (2008). *Review of existing information on the interrelations between soil and climate change.(ClimSoil). Final report* (Technical Report). http://nora.nerc.ac.uk/id/eprint/6452.

Tallis, H., Kareiva, P., Marvier, M., & Chang, A. (2008). An ecosystem services framework to support both practical conservation and economic development. *Proceedings of the National Academy of Sciences, 105*(28), 9457–9464.

TEEB (2010), The Economics of Ecosystems and Biodiversity Ecological and Economic Foundations. Edited by Pushpam Kumar. Earthscan, London and Washington

Ten Brink P., Badura T., Bassi S., Daly, E., Dickie, I., Ding H., Gantioler S., Gerdes, H., Kettunen M., Lago, M., Lang, S., Markandya A., Nunes P.A.L.D., Pieterse, M., Rayment M., Tinch R., (2011). Estimating the Overall Economic Value of the Benefits provided by the Natura 2000 Network. Final Report to the European Commission, DG Environment on Contract ENV.B.2/ SER/2008/0038. Institute for European Environmental Policy / GHK / Ecologic, Brussels 2011

Van Berkel, D. B., & Verburg, P. H. (2014). Spatial quantification and valuation of cultural ecosystem services in an agricultural landscape. *Ecological Indicators, 37*, 163–174.

Väre, H., Lampinen, R., Humphries, C., & Williams, P. (2003). Taxonomic diversity of vascular plants in the European alpine areas. In L. Nagy, G. Grabherr, C. Körner & D. A. Thompson (Eds.), *Alpine biodiversity in Europe* (pp. 133–148). Berlin: Springer.

Verburg, P. H., & Overmars, K. P. (2009). Combining top-down and bottom-up dynamics in land use modeling: Exploring the future of abandoned farmlands in Europe with the Dyna-CLUE model. *Landscape Ecology, 24*(9), 1167–1181. doi:10.1007/s10980-009-9355-7.

Viviroli, D., Dürr, H. H., Messerli, B., Meybeck, M., & Weingartner, R. (2007). Mountains of the world, water towers for humanity: Typology, mapping, and global significance. *Water Resources Research, 43*(7), W07447.

Wild Europe. (2012). *A working definition of European Wilderness and wild areas*. Wild Europe Initiative.

WWF. (2010). *Assessment of the restoration of potential along the Danube and main tributaries*. Vienna: WWF.

Part II
Rewilding and Biodiversity

Chapter 4
Bringing Large Mammals Back: Large Carnivores in Europe

Luigi Boitani and John D. C. Linnell

Abstract The last century has seen a dramatic reversal in the status of large carnivores in Europe. A suite of co-occurring factors has permitted a large-scale recovery of most populations. We currently recognise 10 populations of each species, most of which are transboundary in nature. The sizes of these populations vary from some tens to many thousand, with current estimates being around 17,000 bears, 10,000 wolves and 10,000 lynx in Europe (excluding Russia). As the situation moves from averting extinction to planning recovery it is logical to ask how far the recovery can go, and what our conservation goals should be, especially in light of the emerging rewilding discourse. For a variety of ecological, practical and strategic reasons, it seems unlikely that restoring "wilderness" or "natural ecological processes" (in the sense that human activity and influence are excluded) will serve as general models for large carnivore conservation on a large scale. We suggest a focus on developing a "coexistence" model that aims to create a sustainable interaction between humans and large carnivores by encouraging conservation of these species in very large areas of the European landscape, encouraging the development of a wide range of ecological processes, including predation and scavenging, while accepting that human influence on all trophic levels is pervasive, legitimate, necessary and often even desirable. This constitutes a desire to create a new form of relationship between humans and wildness that has never existed before, and therefore does not fall within the conventional meanings of the rewilding paradigm.

Keywords Large carnivores · Coexistence · Natural ecological process · Herbivory · Social tolerance · Human impact

L. Boitani (✉)
Department of Biology and Biotechnologies, University of Rome "La Sapienza", Viale Università 32, 00185 Rome, Italy
e-mail: luigi.boitani@uniromal.it

J. D. C. Linnell
Norwegian Institute for Nature Research, PO Box 5685 Sluppen, 7485 Trondheim, Norway
e-mail: john.linnell@nina.no

H. M. Pereira, L. M. Navarro (eds.), *Rewilding European Landscapes,*
DOI 10.1007/978-3-319-12039-3_4, © The Author(s) 2015

4.1   Introduction

Large mammals are often regarded as flagship species of wild areas and the paradigm of wilderness untouched by, or at least relatively separated from, human activities (Ray et al. 2005). This is especially true for large carnivores. This view largely stems from the historic processes of direct human persecution and indirect habitat change that gradually reduced their presence in human-dominated landscapes such that they only persisted in the residual areas with little or no human activity. As a consequence, the view of large carnivores as beasts of the wilderness became consolidated, particularly in North America (Boitani 1995). Since Europe is home to more than 500 million people and lacks extensive pristine, uninhabited land areas and large protected areas with spectacular aggregations of large mammals, it might appear to have little to offer for large carnivore conservation. However, nothing could be further from the truth. In the last few decades, changes in the socio-economic settings and people's values concerning nature and biodiversity have paved the way for new opportunities for large carnivores. As the situation develops a new conservation paradigm is slowly emerging based on the premises of coexistence instead of exclusion.

In this essay, we firstly describe the current status and trends of the large carnivores in Europe and examine the main causes of the recent increase in numbers and range. Second, we discuss the available opportunities to sustain the positive trends and the challenges in driving the process toward a new balance between carnivores and human activities. Thirdly, we use the insights coming from large carnivore conservation to offer our views on the social and ecological implications of managing the "rewilding" of Europe. In Europe, five carnivore species have been traditionally considered as "large carnivores", but in this essay we will focus on the three most important ones, the grey wolf (*Canis lupus*), the bear (*Ursus arctos*) and the Eurasian lynx (*Lynx lynx*); the other two, the Iberian lynx (*Lynx pardina*) and the wolverine (*Gulo gulo*) are restricted to small areas, respectively in southern Iberia and northern Fennoscandia, and are associated with very specific management issues.

4.2   Trends in Large Carnivores in Europe

To the Edge of Extinction

Bears, wolves and Eurasian lynx were once widespread across most of the European continent. However, intense persecution, prey extermination and habitat conversion led to their near extermination in the nineteenth and early to mid-twentieth centuries (Breitenmoser 1998; Linnell et al. 2009, 2010). As a result of the eradication efforts, all carnivore populations experienced their smallest population sizes and range contraction during early to mid-twentieth century. The declines were particularly extreme in western, central and northern Europe. Wolves were practically exterminated and relict lynx and bear populations only persisted in parts of Sweden and Finland. In southern Europe, precariously small bear populations persisted in the Cantabrian Mountains, the Pyrenees, the Alps and central Italy. Wolves persisted in parts of the Iberian Peninsula and central Italy. In eastern and south-eastern Europe all species persisted to some extent in the Carpathian and Balkan mountains, but populations were generally very much reduced in both range and density.

Multiple Causes of Recovery

From this nadir, a number of factors have interacted to create the conditions for a continental wide recovery of the species. Many carnivore populations were protected by national and European legislations (Bern Convention of 1982, Habitats Directive of 1992) following significant changes in public opinion towards wildlife conservation, which occurred in many countries around this time. However, it is also interesting to note that much of the early recovery in northern and Eastern Europe occurred within hunting management frameworks, often while the carnivores were being harvested (Swenson et al. 1994). Much of this recovery was long before the ideals of conservation biology had been formulated. By this period there had also been a dramatic recovery of European wild herbivore populations, which had experienced a similar fate as the large carnivores during the nineteenth century. Their recovery during the early and mid-twentieth century had been greatly aided by hunting motivated translocations and the introduction of improved hunting legislation that aimed to manage ungulates for sustainable harvest (Linnell and Zachos 2011). In addition, European forest cover had begun to recover from earlier deforestation, both as a result of forest policies and due to reduced human pressure on the land following large-scale rural—urban migration. This reduced pressure led to both an increase in habitat for predators and prey, and led to a lessening of the human persecution pressure on the carnivores (see Chap. 1). Thus, many positive factors coincided to create a positive ecological and legislative environment for large carnivores to recover, although there was much regional variation in the timing and magnitude of the different processes.

Most of the recovery has been natural. Lynx have naturally recolonized much of Fennoscandia, even expanding into northern areas from where they were historically absent (Linnell et al. 2010). Wolves have naturally recolonized Scandinavia, Finland, France, Switzerland and Germany as well as expanding through much larger areas of Italy, Portugal and Spain (Kaczensky et al. 2013). Dispersing wolves are now appearing in areas like Denmark, the Netherlands, and Austria. Fennoscandian and south-eastern European bear populations have also expanded naturally, although bear expansion is slowed by the intrinsic low rates of female dispersal. Active assistance through reintroduction has played only a minor part in the process. Eurasian lynx were successfully reintroduced to the western Alps, the Jura and Vosges mountains, north-eastern Switzerland, central Germany and central Poland (Linnell et al. 2009). The translocation of bears has successfully taken place in the Italian Alps and, less successfully, in the Pyrenees and in central Austria (Clark et al. 2002). There have been no reintroductions of wolves, although a few individuals have been translocated within Sweden in recent years as part of a genetic reinforcement program.

The Current Status of Populations

Europe's large carnivores are currently distributed among 42 nations, each with unique cultural values for biodiversity and different legal platforms for conservation. This cultural, political, and legal diversity within Europe presents major challenges for the conservation of internationally listed species, which often exist in transboundary populations that fall across several international jurisdictions. Management fragmentation is made worse by the fact that many European countries (e.g. Austria, Spain, Germany) are federal countries where responsibility for nature conservation has been decentralised to many sub-national jurisdictions. Large carnivores have all the characteristics of species that are difficult to manage at the scale of Europe's small administrative units: they live at low densities (typically less than $3/100$ km^2), have home range size up to 1000 km^2 and dispersal distances of more than 1000 km (Linnell and Boitani 2012).

In an attempt to facilitate carnivore management at the appropriate scale of biologically meaningful units instead of administrative compartments, the European Commission approved a set of "Guidelines for population level management plans" (Linnell et al. 2008) and identified the main populations across the continent. The populations were identified based on several criteria such as the discontinuity in distribution, geographic features, the species' dispersal distance and the ecological and management contexts. Out of 30 populations (see below), only four occur within a single country and some span up to eight countries. Kaczensky et al. (2013) recently reviewed the conservation status of the European large carnivores in 2012 using data collected by a network of experts across Europe. The following sections are drawn from their report.

Bears

The total number of brown bears in Europe is estimated to be about 17,000 individuals. They occur in 22 countries and 10 main populations (Fig. 4.1): Scandinavian, Karelian, Baltic, Carpathian, Dinaric- Pindos, Eastern Balkan, Alpine, Central Apennine, Cantabrian, and Pyrenean. The largest population is the Carpathian population (>7000 bears), followed by the Scandinavian and Dinaric- Pindos populations (>3000 bears). The other populations are much smaller ranging from several hundred (e.g. Karelian c. 850, Baltic c. 700, Cantabrian c. 200) to less than a hundred (e.g. Central Apennine 40–70, Alps 45–50, Pyrenean 22–27). Only two

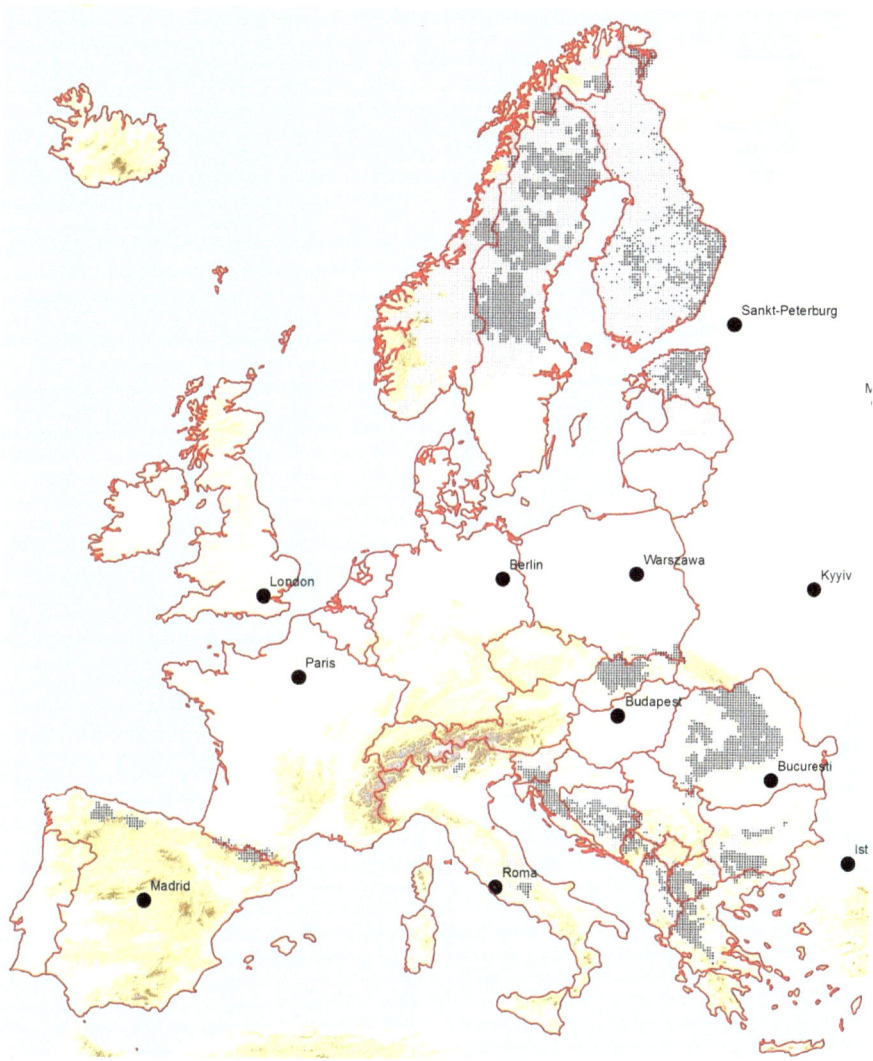

Fig. 4.1 Distribution of bears and their populations in Europe in 2012. *Dark cells* permanent occurrence, *Grey cells* sporadic occurrence. (From Kaczensky et al 2013)

small populations (Alpine and Pyrenean) have been reinforced with animals translocated from Slovenia.

Trends in number and range expansion are generally positive: all populations are either stable in number or show an increase (Scandinavian, Karelian, Dinaric-Pindos, Baltic, Cantabrian, and Pyrenean); their range is also stable or slightly expanding. With the exception of the small populations of the Cantabrian, Central Apennine and Pyrenees, no populations are threatened and most of them are well protected by effective legislation that severely limits human-induced mortality. The Habitats Directive provides full protection for all bears in the European Union under Annex IV, although moderate culling is allowed under article 16 derogations in Sweden, Finland, Romania, Estonia, Bulgaria, Slovenia and Slovakia. Overall, the level of conflict with human activity is surprisingly low for such an opportunistic species that feeds on a large variety of items. With the notable exception of Norway, Spain and Slovenia, all other countries pay small amounts in compensation for bear damages to livestock and other agricultural products. The overall cost of compensation in Europe is in the order of 3 million € per year (Kaczensky et al. 2013). In spite of their size and potential for being dangerous to human lives, bears in Europe are not a significant threat to humans and injuries or lethal attacks are limited to a few occasional cases.

Wolves

There are probably more than 10,000 wolves in Europe. They occur in all countries except the island states (Ireland, Iceland, United Kingdom, Cyprus and Malta) and the Benelux countries. At least 10 main wolf populations can be identified: north-western Iberian, Sierra Morena (southern Spain), Alpine, Italian Peninsula, Carpathian, Dinaric-Balkan, Baltic, Karelian, Scandinavian and Central European Lowlands (Fig. 4.2). The largest populations are in southern and Eastern Europe such as the Carpathian and the Dinaric- Balkan populations (>3000 wolves each), followed by the north-western Iberian (~2500 wolves) and the Baltic (>1000 wolves). Other populations are an order of magnitude smaller (numbering in the low hundreds with the Italian Peninsula population being somewhat larger, in the range of 600–800 wolves) and the Sierra Morena population in southern Spain now reduced to just one pack detected in 2012. No wolf reintroductions (i.e. release of individuals where the species had been exterminated in historical times) have ever been carried out in Europe, although most recently there have been a few translocations of individuals within wolf range inside Sweden.

Trends in numbers and range size are generally positive since the last estimates in 2005. With the exception of Sierra Morena population, all populations are either stable or increasing and there is good evidence of large dispersal movements potentially re-connecting populations, such as the Alpine and Dinaric or the Scandinavian and Karelian. However, some countries have seen their national estimates decreasing such as Albania, Finland, Macedonia, and Portugal for the subpopulation south of the Douro River, where the social, ecological and political conditions for wolf acceptance have significantly deteriorated recently.

Most European wolves are covered by the full protection offered (with derogations possible under article 16) by Annexes II (requires establishment of Natura 2000 sites) and IV (strict protection) of the Habitats Directive although there are several exceptions of countries that have their wolf populations (or just part of it) in Annex V (which permits regulated harvest): for example, the Baltic countries,

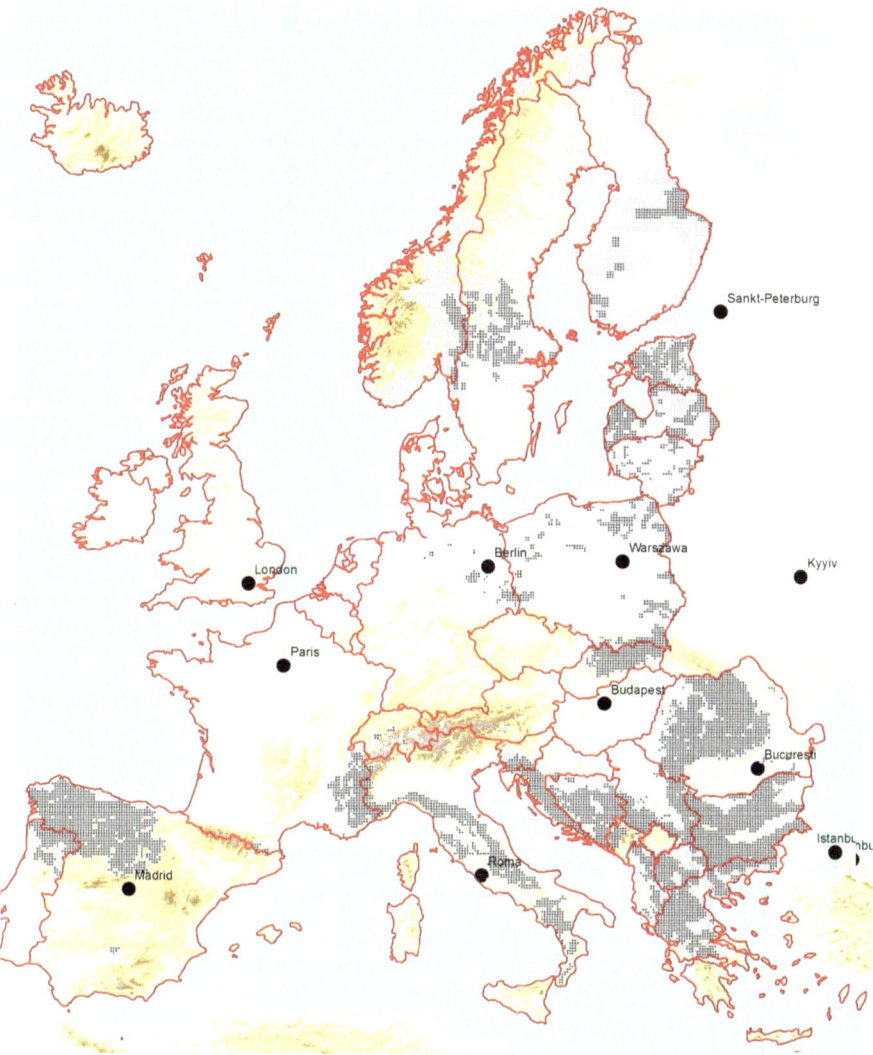

Fig. 4.2 Distribution of wolves and their populations in Europe in 2012. *Dark cells* permanent occurrence, *Grey cells* sporadic occurrence. (From Kaczensky et al 2013)

Bulgaria, Poland, Slovakia, parts of Greece and parts of Finland. Depredation by wolves on livestock is one of the most ancient conflicts that humans have sustained against wildlife and it is still widespread across Europe. The total economic loss is estimated to be in the range of 8 million € and about 20,000 domestic animals, mostly sheep, are killed annually with huge variations between countries. The costs of adopting damage prevention measures can also be significant, and in some countries is far greater than the cost of damage prevention. In addition to the economic and material costs of livestock depredation, many hunters perceive wolves as competitors for shared game. Moreover, wolves have generated a wide range of, often intense, social and political conflicts in western and northern Europe, as they have become political symbols for many social issues including urban vs. rural and modern vs. traditional tensions. Although historical evidence indicates that wolf attacks on humans were widespread in the past, there have only been a handful of exceptional cases detected during the last century (Linnell et al. 2002).

Eurasian Lynx

The total number of lynx in Europe is estimated to be 9000-10,000 individuals. They occur in 23 countries divided into 10 main populations (Fig. 4.3): five of these ten populations are autochthonous (Scandinavian, Karelian, Baltic, Carpathian and Balkan), the other populations stem from reintroductions in the 1970s and 1980s (Dinaric, Alpine, Jura, Vosges-Palatinian and Bohemian- Bavarian) (Linnell et al. 2009). Of the autochthonous populations, only the Balkan one is of conservation concern, having been reduced to about 40–50 individuals and showing no signs of significant recovery. The reintroduced populations are all small in the range of 20 individuals in the Vosges-Palatinian to about 150 in the Alpine population. In addition, lynx roam the Harz Mountains of central Germany because of recent reintroductions.

The general trend in numbers is stable or slightly increasing, although there is some concern for the long-term viability of the reintroduced populations due to small population effects and the risk of inbreeding. Most of the lynx populations are strictly protected and derogations under article 16 of the Habitats Directive are used to harvest the populations in Sweden, Latvia and Finland. Estonia is unique within the EU having the lynx on annex V, which permits regulated harvest as a game species. Large conflicts with livestock owners are limited to the northern populations. The only country with a major conflict with sheep is Norway, where about 7000–10,000 sheep are compensated annually. In addition, thousands of semi-domestic reindeer deaths are attributed to lynx depredation annually in Norway, Sweden and Finland. Elsewhere the level of livestock depredation is very small. However, the level of conflict with hunters is widespread across Europe who perceive lynx as a competitor for wild ungulates, especially roe deer (Breitenmoser et al. 2010).

4.3   How Far Can We Take the Recovery Process?

Although the status of large carnivores across Europe is heterogeneous and dynamic there are grounds for increasing (though still cautious) optimism concerning their future status. Apart from a few small populations that are clearly still threatened (such as bears in central Italy and the Pyrenees), the main task for the future is more one of sustaining their recovery than of saving them from extinction (Swenson et al.

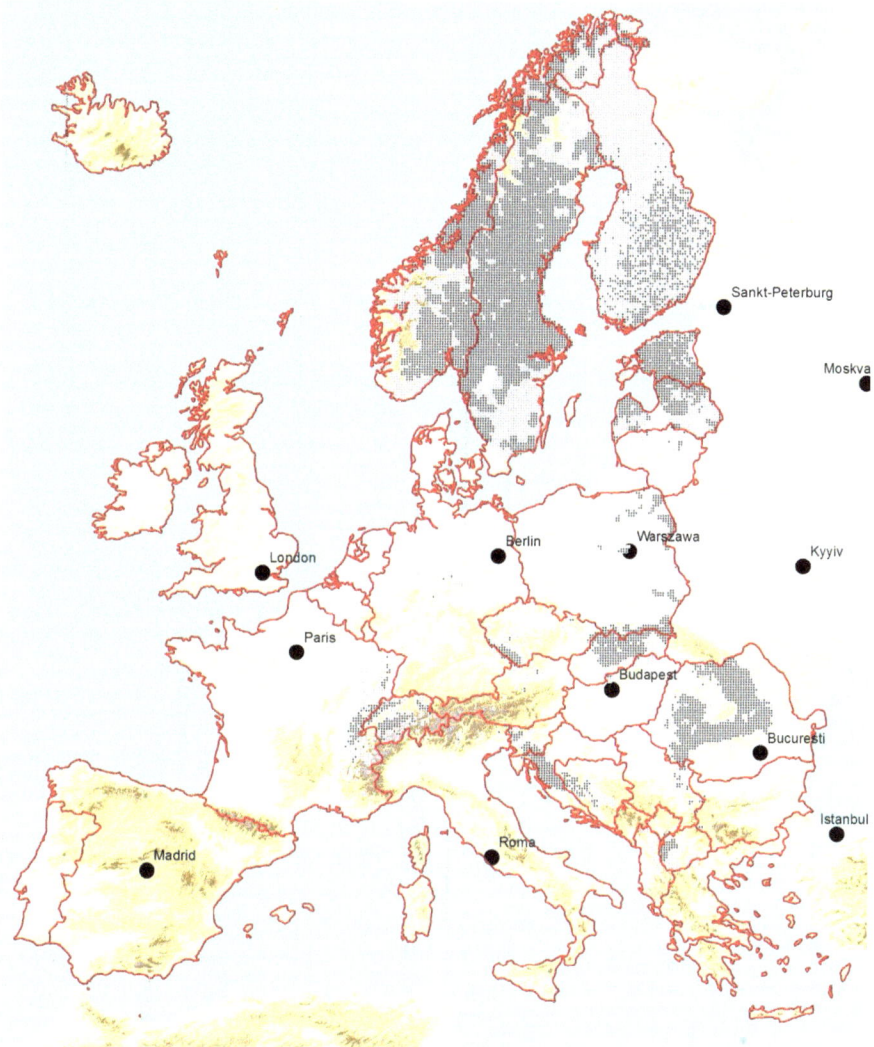

Fig. 4.3 Distribution of lynx and their populations in Europe in 2012. *Dark cells* permanent occurrence, *Grey cells* sporadic occurrence. (From Kaczensky et al 2013)

1998). This leads to asking how far the recovery may go. In other words, what level of conservation ambition should we hope for (Linnell et al. 2005)? While the short-term recovery goals have always been about achieving "population viability" to safeguard against population and species extinctions, the present conservation discourse is now increasingly moving to one of "ecological functionality". Although the rewilding movement offers many diverse points of view (Donlan et al. 2006; see Chaps. 1 and 9), it has been placing a lot of emphasis on the restoration of "natural processes" with the often stated understanding that this excludes human activity.

So far these ideas have been mainly directed towards restoring herbivory, although they have met with considerable debate (Hodder et al. 2005; Kirby 2009). There is an increasing trend to also extend them to predation and we feel it is also important to raise some questions concerning whether it is possible, or even desirable, to restore "natural predation processes" (Andersen et al. 2006). It is important to ask what this term means, and even ask if it should be the benchmark goal for large carnivore conservation strategies from strategic and value based points of view.

What are the Characteristics of "Natural Predation Processes"?

Describing the nature of predator—prey dynamics for large carnivores and large herbivores has been an ongoing theme in ecological research for the last 60–70 years at least. From the early writings of Aldo Leopold through to the on-going long-term predator-prey studies of wolf—moose relationships on Isle Royale and other parts of North America and Africa, there has been much speculation about the relative importance of top-down and bottom-up factors in regulating densities of predators and prey (e.g. Skogland 1991; Mech and Peterson 2003). The discussion has also spilled over into discussions in modern conservation biology about the pervasiveness of trophic cascades and the role of predators as ecological keystones (Ray et al. 2005; Terborgh and Estes 2010). Reviewing this vast literature is beyond the scope of this chapter, but it is possible to extract some findings relevant for our discussion. Firstly, predators have been shown to have a diversity of behavioural and demographic effects on their prey and on other aspects of ecosystem function through trophic cascades. However, the strength of these impacts varies considerably across space (Melis et al. 2009, 2010) and time (Mech and Peterson 2003) and with the behaviour of the predator and the anti-predator strategy of the prey. Secondly, the impact of predators on prey is very much dependent on the numerical response of predator density as well on the functional responses of kill rates to changes in prey density (Andersen et al. 2006). Moreover, large carnivore populations operate at very large scales, with home ranges spreading across hundreds of square kilometres (Nilsen et al. 2005) and dispersal distances covering hundreds of kilometres (Samelius et al. 2012). This implies that spatial dynamics of large carnivores can only be measured on such large scales, making it hard to predict impacts at local scales. Additionally, many fine-scaled factors such as variation in habitat structure or snow depth can introduce micro-level modifications to the larger scale processes (Gorini et al. 2012), which introduce uncertainties in predicting larger scale dynamics. Furthermore, both large carnivores and large herbivores are influenced by external factors such as climate and disease that have the potential to induce dramatic changes in population sizes and predator prey relationships. Finally the empirical data underpinning our understanding of predator-prey systems is very limited, especially for systems with multiple predators and multiple prey, and time series are almost entirely rather short.

The present state of knowledge is sufficient to have a good qualitative idea of the impacts of predation and the types of predator-prey dynamics that can occur. However, it is very hard to predict in a quantitative way what will occur in any given location. This is especially true for Europe, where there have been very few long-term predator-prey studies. Therefore, it is rather difficult to speculate about what "natural predation processes" will actually be in any given location.

The Pervasive Impact of Humans

Despite the existence of many large protected areas where human impacts are minimised, just about all predator-prey systems on earth are impacted by humans in various ways. The most obvious and immediate effect is through human induced mortality of both predators and prey. While some of the planet's largest protected areas may insulate some large herbivores from human exploitation, there is still pervasive human impact through poaching and legal harvest within protected areas or on the herbivores that seasonally migrate outside the borders. For large carnivores, the situation is even worse as their wide-ranging movements more often carry them beyond protected area borders (Woodroffe and Ginsberg 1998). In a European context, where protected areas are often small, there are probably very few large predator individuals, let alone populations, that live their lives entirely inside protected areas (Linnell et al. 2001).

In addition to the deliberate targeting of these species, there are many other sources of mortality which humans induce, such as through vehicle collisions (Langbein et al. 2011) and cases where disease is transferred from domestic to wild species. Furthermore, humans have very strong impacts on herbivores through their manipulation of habitats (see Chap. 8). Forestry and agricultural practices have dramatic impacts on vegetation structure and productivity that can have both positive and negative impacts on herbivore and carnivore populations (e.g. Gill et al. 1996; Torres et al. 2011). In general, small scale forestry and agriculture lead to situations that increase productivity and benefit many herbivores. The winter feeding of wild herbivores is a widespread activity across most of northern, eastern and central Europe which has the potential to greatly influence herbivore distribution and density (Putman et al. 2011). Long-term deposition of nitrogen and climate change can also have dramatic impacts on the productivity of vegetation (Holland et al. 2005). Another impact comes from competition between domestic and wild species. Domestic herbivore densities tend to exceed those of wild herbivores and can have dramatic impacts on habitat structure and the productivity of vegetation, as well as providing potential prey items for predators. In many areas, animals of domestic origin have been, and still are, critical prey for large carnivores (Mattisson et al. 2011; Peterson and Ciucci 2003). Even predators of domestic origin (domestic cats and dogs) can compete with wild predators. A final impact occurs through the behavioural disturbance that human presence and activity can induce in both predators and prey (Moen et al. 2012). Given the mobility of both large carnivores and large herbivores, the spatial impacts of these diverse perturbations are likely to influence the structure and functions of populations on scales of at least tens and hundreds of kilometres. Across Europe, there is a very high degree of diversity in the ways habitats, herbivores and carnivores are managed, such that actions in neighbouring countries could well have dramatic impacts on predator-prey dynamics even beyond their own borders (Putman et al. 2011; Kaczensky et al. 2013; Linnell and Boitani 2012).

Despite the pervasive impacts of humans, the recent history of large carnivore and large herbivore recovery in Europe has shown that these species have a remarkable ability to persist and thrive in human-modified landscapes. There are clear species-specific differences in this tolerance, with wolves and roe deer for example being especially tolerant of modified landscapes, and species such as bears and wild reindeer being least tolerant. Certainly there are limits to tolerance. Extreme habitat modification for intensive agriculture and high rates of disturbance can make many areas unliveable for many species (e.g. Schadt et al. 2002; Güthlin et al. 2011; Jedrzejewski et al. 2008). A lot of transport infrastructure has the potential to create barriers (Kaczensky et al. 2003). However, in general none of these species require areas free of human intervention, and most will in fact benefit to some extent from many low-intensity human activities (Basille et al 2009; Torres et al. 2011).

The Social Tolerance of Humans for Large Carnivores and Large Herbivores

Despite the potential for carnivores and herbivores to persist and even achieve very high density in human modified landscapes, the major limit to the densities they achieve is likely to be set by human tolerance for their presence. Herbivores create a diversity of conflicts with humans, ranging from damage to crops and forestry, the transfer of disease to domestic animals, and vehicle collisions (Gordon 2009). Regardless of the real level of conflict, large carnivores are associated with conflicts such as depredation on livestock, destruction of beehives, and competition with hunters for shared game. The level of social and political conflict that results from efforts to conserve species such as wolves and bears can be intense in some areas, especially in places where they return after long absences (Benhammou and Mermet 2003; Skogen et al. 2006). The effect of these conflicts is largely to reduce human tolerance for the presence of these species, which tends to result in efforts to limit the density or distribution of these species through lethal means.

The Problem of Natural Processes as a Goal

Based on the arguments presented above there are clearly some problems with having a "return to natural processes" as an ecological objective for large carnivores and large herbivores in Europe. Firstly, we do not exactly know what these processes look like; making it hard to recognise the state even if we could reach it. Historical analysis represents very little help seeing as humans have been severely affecting all trophic levels in Europe for many millennia. Secondly, the impacts of humans on habitats, herbivores and carnivores is so pervasive that there simply are no areas large enough in Europe for these processes to occur without there being a major impact of human activity on all trophic levels. Thirdly, because of the conflicts that both herbivores and carnivores can induce with human activities there is likely to be little acceptance for allowing their populations to develop without some form of intervention and control (both in terms of reinforcement and reduction of populations under varying contexts)—which in turn is likely to impact the dynamics between predators and prey.

In other words, it is hard to know what these natural processes look like, it will be hard to achieve them in practice, and the process of trying to achieve them may be associated with significant levels of material and social conflict. Combined, these arguments represent severe technical and strategic obstacles for any effort to pursue "natural ecological processes" (in the sense that they are free from human interference) within a wilderness setting as a conservation goal for a large herbivore-large carnivore predator prey system in a European context. Another fundamental issue concerns the implicit assumption of these "natural process" goals that humans are not part of nature, and that their interactions with nature are not natural. This assumption has been instrumental in the construction of the "wilderness" ideal (Cronon 1995; Marris 2011; see Chap. 2). This dualistic worldview has been heavily rejected in recent years by anthropologists and nature philosophers in favour of a much more integrated view that firmly places humans as integral and interactive parts of nature (Descola and Pålsson 1996). Following this emerging line of argument, the interactions between humans and nature should be as much a legitimate target of conservation as the interactions between non-human parts of nature.

From Wilderness and Natural Processes to a Future Orientated Coexistence

Our arguments so far have caused us to raise serious questions about the extent to which "natural ecological processes" or "wilderness" are either potentially achievable or even desirable goals for the general conservation of large carnivores and large herbivores. Therefore, the question remains: what we should replace it with? The recent history of carnivore and herbivore conservation in Europe and their current status show that we have an incredible opportunity to integrate these species into very large areas of the European landscape. In many areas we may well be able to restore the full assemblage of species that have been found on the continent for the last few millennia. In some few areas this may occur in areas where there has been little human modification of habitat and where there is minimal direct influence by humans on the species. However, these areas will be the exception. As we have seen most of these species are tolerant to many forms of human activity. In principle there are very few parts of the continent where at least some of the large herbivores and carnivores will not be able to live. This implies that wolves, bears, lynx, bison, moose, red deer, roe deer, ibex, chamois, wild boar and other species can look forward to rather wide distributions in the coming decades. The fact that this conservation will be occurring in multi-use landscapes implies that all trophic levels and interactions will be, to some extent, influenced by humans, often in radical ways. Despite this modification there is a huge potential for a diversity of ecological processes to resume, including predation and scavenging, albeit in modified ways.

In other words, even if "wilderness" is unattainable there is a huge scope for increasing the amount of "wild" in most parts of the European landscape. This conservation view is best termed the "coexistence" approach as it seeks to integrate wildlife and humans in a shared landscape. Its focus on achievable "wildness", rather than unobtainable "wilderness", allows for a much more optimistic view of conservation, where every small recovery can be viewed as a success, rather than lamenting how much it falls short of some ideal (Kirby 2009; Marris 2011). A wolf raising pups in a Spanish agricultural plain is a triumph for coexistence as it shows the dramatic return of a degree of wildness to an otherwise heavily domesticated landscape, even if the functionality of the system is as far from wilderness as you could ever imagine. Having wild animals back in more parts of the landscape will also secure a far greater degree of long term viability (for example by increasing connectivity) than would be achieved from having some few very special wilderness areas, even if they could be obtained. Conserving large carnivores only in some small "wilderness" areas is simply impossible (i.e. in a land sparing approach sensu Phalan et al. 2011) because of their spatial needs (Linnell and Boitani 2012).

The coexistence approach represents many challenges as it does increase the area of interface between humans and wildlife, which potentially opens for more conflicts (Gordon 2009; Linnell 2013). However, the approach also opens for humans to enter into mindful and interactive relationships with the wildlife and allows them to mitigate or react to these issues, and find some form of dynamic and active relationship with the species that share their landscapes. For many cases this relationship may require re-adopting some traditional practices (for example shepherding methods), but it will also require adopting many new and innovative practices such as electric fences and green bridges. While the coexistence approach is in many ways trying to take advantage of changing situations in Europe (see Chap. 1) it has nothing retrospective about it, which makes it stand out from many interpretations of "rewilding" (where the "re" suffix implicitly suggests a retrospective component). Rather it is a future orientated approach that seeks to build a sustainable relationship with wildlife in shared landscapes. This has rarely been achieved before in our history, and certainly has not been attempted on a continental scale in modern times, with all the pressures that our modern society is placing on the land. It will be an essentially hands on approach, requiring a lot of adaptive management as it seeks to find a way forward that can adjust to the ecological and societal dynamics of the human and non-human actors that are trying to share the same landscapes.

4.4 Conclusions

The last 40–50 years have seen a dramatic reversal in fortune for many large carnivore and large herbivore species. When the focus moves beyond saving them from extinction it is logical to begin exploring long-term conservation goals. It may be possible to create some areas with full species assemblages and a minimum of direct human interference on the species and their habitat (see Chap. 9). Such areas are clearly of a high degree of conservation and scientific interest. However, the spatial scales at which the dynamics of large carnivores and large herbivores occur and the huge human pressure on space and resources in Europe will inevitably lead to a range of subtle human influences on these areas, and will prevent these approaches from having general value at large scales. Therefore, we believe that defining goals in terms of "wilderness" and "natural ecological processes" (in ways that exclude humans and human activities) has very little relevance as a general model for large carnivore conservation in Europe. In contrast, because all these species have shown a high degree of tolerance for many human activities it is possible to imagine a future based on "coexistence" where they are integrated into a very large proportion of the wider multi-use landscape. This will permit a large degree of wildness to appear in many areas. The challenge will not be to minimise human impacts on them, but to find ways for these interactions to occur in a sustainable manner. This future will have fallen outside many of the conventional "rewilding" philosophies, which often have retrospective and hands-off connotations. It is a state that has rarely been achieved before and will require constant management and intervention. Within this framework, there is enormous scope for creating a "new-wild" which is built on such key ideas as diversity, interaction, tolerance, sustainability, and coexistence.

References

Andersen, R., Linnell, J. D. C. & Solberg, E. J. (2006). The future role of large carnivores on terrestrial trophic interactions: The northern temperate view. In K. Danell, R. Bergström, P. Duncan, & J. Pastor (Eds.), *Large herbivore ecology, ecosystem dynamics and conservation* (pp. 413–448). Cambridge: Cambridge University Press.

Basille, M., Herfindal, I., Santin-Janin, H., Linnell, J. D. C., Odden, J., Andersen, R., Høgda, K. A., & Gaillard, J. M. (2009). What shapes Eurasian lynx distribution in human dominated landscapes: Selecting prey or avoiding people? *Ecography, 32,* 683–691.

Benhammou, F., & Mermet, L. (2003). Strategy and geopolitics of the opposition against the preservation of nature: The bear in the Pyrenees. *Natures Sciences Sociétes, 11,* 381–393.

Boitani, L. (1995). Ecological and cultural diversities in the evolution of wolf human relationships. In L. N. Carbyn, S. H. Fritts, & D. R. Seip (Eds.), *Ecology and conservation of wolves in a changing world* (pp. 3–12). Alberta: Canadian Circumpolar Institute.

Breitenmoser, U. (1998). Large predators in the Alps: The fall and rise of man's competitors. *Biological Conservation, 83,* 279–289.

Breitenmoser, U., Ryser, A., Molinari-Jobin, A., Zimmermann, F., Haller, H., Molinari, P., & Breitenmoser-Würsten, C. (2010). The changing impact of predation as a source of conflict between hunters and reintroduced lynx in Switzerland. In D. W. Macdonald & A. J. Loveridge (Eds.), *Biology and conservation of wild felids* (pp. 493–506). Oxford: Oxford University Press.

Clark, J. D., Huber, D., & Servheen, C. (2002). Bear reintroductions: Lessons and challenges. *Ursus, 13*, 335–345.

Cronon, W. (1995). *Uncommon ground: Rethinking the human place in nature.* New York: W W Norton & Co.

Descola, P., & Pålsson, G. (1996). *Nature and society: Anthropological perspectives.* London: Routledge.

Donlan, C. J., Berger, J., Bock, C. E., Bock, J. H., Burney, D. A., Estes, J. A., Foreman, D., Martin, P. S., Roemer, G. W., Smith, F. A., Soulé, M., & Greene, H. W. (2006). Pleistocene rewilding: An optimistic agenda for twenty-first century conservation. *American Naturalist, 168,* 660–681.

Gill, R. M. A., Johnson, A. L., Francis, A., Hiscocks, K., & Peace, A. J. (1996). Changes in roe deer (*Capreolus capreolus* L.) population density in response to forest habitat succession. *Forest Ecology and Management, 88,* 31–41.

Gordon, I. J. (2009). What is the future for wild, large herbivores in human-modified agricultural landscapes. *Wildlife Biology, 15,* 1–9.

Gorini, L., Linnell, J. D. C., May, R., Panzacchi, M., Boitani, L., Odden, M., & Nilsen, E. B. (2012). Habitat heterogeneity and mammalian predator-prey interactions. *Mammal Review, 42,* 55–77.

Güthlin, D., Knauer, F., Kneib, T., Küchenhoff, H., Kaczensky, P., Rauer, G., Jonozovic, M., Mustoni, A., & Jerina, K. (2011). Estimating habitat suitability and potential population size for brown bears in the eastern Alps. *Biological Conservation, 144,* 1733–1741.

Hodder, K. H., Bullock, J. M., Buckland, P. C., & Kirby, K. J. (2005) Large herbivores in the wildwood and modern naturalistic grazing systems. Peterborough: English Nature, (ISSN 0967-876X).

Holland, E. A., Braswell, B. H., Sulzman, J., & Lamarque, J. F. (2005). Nitrogen deposition onto the United States and western Europe: Synthesis of observations and models. *Ecological Applications, 15,* 38–57.

Jedrzejewski, W., Jedrzejewska, B., Zawadzka, B., Borowik, T., Nowak, S., & Myslajek, R. (2008). Habitat suitability model for Polish wolves based on long term national census. *Animal Conservation, 11,* 377–390.

Kaczensky, P., Knauer, F., Krze, B., Jonozovic, M., Adamic, M., & Grossow, H. (2003). The impact of high speed, high volume traffic axes on brown bears in Slovenia. *Biological Conservation, 111,* 191–204.

Kaczensky, P., Chapron, G., Von Arx, M., Huber, D., Andrén, H., & Linnell, J. (2013). *Status, management and distribution of large carnivores—bear, lynx, wolf and wolverine—in Europe.* Rome: Istituto di Ecologia Applicata.

Kirby, K. J. (2009). Policy in or for the wilderness? *British Wildlife, 20,* 59–62.

Langbein, J., Putman, R., & Pokorny, B. (2011). Traffic collisions involving deer and other ungulates in Europe and available measures for mitigation. In R. Putman, M. Apollonio, & R. Andersen (Eds.), *Ungulate management in Europe: Problems and practices* (pp. 215–259). Cambridge: Cambridge University Press.

Linnell, J. D. C. (2013). *From conflict to coexistence: Insights from multi-disciplinary research into the relationships between people, large carnivores and institutions.* Rome: Istituto di Ecologia Applicata.

Linnell, J. D. C., & Boitani, L. (2012). Building biological realism into wolf management policy: The development of the population approach in Europe. *Hystrix—Italian Journal of Mammalogy, 23,* 80–91.

Linnell, J. D. C., & Zachos, F. E. (2011). Status and distribution patterns of European ungulates: Genetics, population history and conservation. In R. Putman, M. Apollonio, & R. Andersen (Eds.), *Ungulate management in Europe: Problems and practices* (pp. 12–53). Cambridge: Cambridge University Press.

Linnell, J. D. C., Andersen, R., Kvam, T., Andrén, H., Liberg, O., Odden, J., & Moa, P. (2001). Home range size and choice of management strategy for lynx in Scandinavia. *Environmental Management, 27,* 869–879.

Linnell, J. D. C., Løe, J., Okarma, H., Blancos, J. C., Andersone, Z., Valdmann, H., Balciauskas, L., Promberger, C., Brainerd, S., Wabakken, P., Kojola, I., Andersen, R., Liberg, O., Sand, H., Solberg, E. J., Pedersen, H. C., Boitani, L., & Breitenmoser, U. (2002). The fear of wolves: A review of wolf attacks on humans. *Norwegian Institute for Nature Research Oppdragsmelding, 731,* 1–65.

Linnell, J. D. C., Promberger, C., Boitani, L., Swenson, J. E., Breitenmoser, U., & Andersen, R. (2005). The linkage between conservation strategies for large carnivores and biodiversity: The view from the "half-full" forests of Europe. In J. C. Ray, K. H. Redford, R. S. Steneck, & J. Berger (Eds.), *Carnivorous animals and biodiversity: Does conserving one save the other?* (pp. 381–398). Washington, DC: Island Press.

Linnell, J. D. C., Salvatori, V., & Boitani, L. (2008). Guidelines for population level management plans for large carnivores in Europe. A large carnivore initiative for Europe report prepared for the European Commission (contract 070501/2005/424162/MAR/B2).

Linnell, J. D. C., Breitenmoser, U., Breitenmoser-Würsten, C., Odden, J., & von Arx, M. (2009). Recovery of Eurasian lynx in Europe: What part has reintroduction played? In M. W. Hayward & M. J. Somers (Eds.), *Reintroduction of top-order predators* (pp. 72–91). Oxford: Wiley-Blackwell.

Linnell, J. D. C., Brøseth, H., Odden, J., & Nilsen, E. B. (2010). Sustainably harvesting a large carnivore? Development of Eurasian lynx populations in Norway during 160 years of shifting policy. *Environmental Management, 45,* 1142–1154.

Marris, E. (2011). *Rambunctious garden: Saving nature in a post-wild world.* New York: Bloomsbury Publishing USA.

Mattisson, J., Odden, J., Nilsen, E. B., Linnell, J. D. C., Persson, J., & Andrén, H. (2011). Factors affecting Eurasian lynx kill rates on semi-domestic reindeer in northern Scandinavia: Can ecological research contribute to the development of a fair compensation system? *Biological Conservation, 144,* 3009–3017.

Mech, L. D., & Peterson, R. O. (2003). Wolf-prey relations. In L. D. Mech & L. Boitani (Eds.), *Wolves: Behavior, ecology, and conservation* (pp. 131–160). Chicago: University of Chicago Press.

Melis, C., Jedrzejewska, B., Apollonio, M., Barton, K., Jedrzejewski, W., Linnell, J. D. C., Kojola, I., Kusak, J., Adamic, M., Ciuti, S., Delehan, I., Dykyy, I., Krapine, K., Mattioli, L., Sagaydak, A., Samchuk, N., Schmidt, K., Shkvyrya, M., Sidorovich, V. E., Zawadzka, B., & Zhyla, S. (2009). Predation has a greater impact in less productive environments: Variation in roe deer (*Capreolus capreolus*) population density across Europe. *Global Ecology and Biogeography, 18,* 724–734.

Melis, C., Basille, M., Herfindal, I., Linnell, J. D. C., Odden, J., Gaillard, J. M., Høgda, K. A., & Andersen, R. (2010). Roe deer population growth and lynx predation along a gradient of environmental productivity and climate in Norway. *Ecoscience, 17,* 166–174.

Moen, G. K., Støen, O. G., Sahlén, V., & Swenson, J. E. (2012). Behaviour of solitary adult Scandinavian brown bears (*Ursus arctos*) when approached by humans on foot. *Plos ONE, 7,* e31699.

Nilsen, E. B., Herfindal, I., & Linnell, J. D. C. (2005). Can intra-specific variation in carnivore home-range size be explained using remote sensing estimates of environmental productivity? *Ecoscience, 12,* 68–75.

Peterson, R. O., & Ciucci, P. (2003). The wolf as a carnivore. In L. D. Mech & L. Boitani (Eds.), *Wolves: Behavior, ecology, and conservation* (pp. 104–130). Chicago: University of Chicago Press.

Phalan, B., Onial, M., Balmford, A., & Green, R. E. (2011). Reconciling food production and biodiversity conservation: Land sharing and land sparing compared. *Science, 333,* 1289–1291.

Putman, R., Apollonio, M., & Andersen, R. (Eds.). (2011). *Ungulate management in Europe: Problems and practices.* Cambridge: Cambridge University Press.

Ray, J. C., Redford, K. H., Steneck, R. S., & Berger, J. (2005). *Large carnivores and the conservation of biodiversity.* Washington, DC: Island Press.

Samelius, G., Andrén, H., Liberg, O., Linnell, J. D. C., Odden, J., Ahlqvist, P., Segerström, P., & Sköld, K. (2012). Spatial and temporal variation in natal dispersal by Eurasian lynx in Scandinavia. *Journal of Zoology, 286,* 120–130.

Schadt, S., Revilla, E., Wiegand, T., Knauer, F., Kaczensky, P., Breitenmoser, U., Bufka, L., Cerveny, J., Koubek, P., Huber, T., Stanisa, C., & Trepl, L. (2002). Assessing the suitability of central European landscapes for the reintroduction of Eurasian lynx. *Journal of Applied Ecology, 39,* 189–203.

Skogen, K., Mauz, I., & Krange, O. (2006). Wolves and Eco-power. A French-Norwegian analysis of the narratives of the return of large carnivores. *Journal of Alpine Research, 94,* 78–87.

Skogland, T. (1991). What are the effects of predators on large ungulate populations? *Oikos, 61,* 401–411.

Swenson, J. E., Sandegren, F., Bjärvall, A., Söderberg, A., Wabakken, P., & Franzén, R. (1994). Size, trend, distribution and conservation of the brown bear *Ursus arctos* population in Sweden. *Biological Conservation, 70,* 9–17.

Swenson, J. E., Sandegren, F., Bjärvall, A., & Wabakken, P. (1998). Living with success: Research needs for an expanding brown bear population. *Ursus, Internatkional Conference on Bear Reserach and Management, 10,* 17–23.

Terborgh, J., & Estes, J. A. (2010). *Trophic cascades: Predators, prey, and the changing dynamics of nature.* Washington, DC: Island Press.

Torres, R. T., Carvalho, J. C., Panzacchi, M., Linnell, J. D. C., & Fonseca, C. (2011). Comparative use of forest habitats by roe deer and moose in a human-modified landscape in south-eastern Norway during winter. *Ecological Research, 26,* 781–789.

Woodroffe, R., & Ginsberg, J. R. (1998). Edge effects and the extinction of populations inside protected areas. *Science, 280,* 2126–2128.

Chapter 5
Top Scavengers in a Wilder Europe

Ainara Cortés-Avizanda, José A. Donázar and Henrique M. Pereira

Abstract The concept of rewilding should not only be applied to recovering habitat and vertebrate populations but also to the restoration of complex ecological processes. Large avian scavengers are the target of restoration programs including conservation measures linked to the manipulation of food resources but we lack of a general approach to understanding how scavengers and the ecosystem services they provide will fit into a rewilded Europe. Carcasses play an important role in ecosystem functioning and in the energy flux within food webs. Large ungulates carcases availability, in particular, has, through the course of evolution, given way to the appearance of "true" scavenger strategies, displayed by large body-sized avian organisms (vultures) whose guilds are structured by complex interspecific relationships. Yet, livestock raised in traditional agro-grazing systems have historically replaced wild ungulates as the main food source for vultures. More recently, modern farm intensification, stricter European Union legislation that banned the abandonment of carcasses, and increasing human-vulture conflicts contributed to plunging vulture populations, leading to an unprecedented crisis. Consequently, supplementary feeding became a management tool used worldwide to aid in the recovery of their decimated populations. These so-called vulture restaurants,

A. Cortés-Avizanda (✉) · H. M. Pereira
Centro de Biologia Ambiental, Faculdade de Ciências da Universidade de Lisboa
Campo Grande, 1749-016, Lisboa, Portugal
e-mail: cortesavizanda@gmail.com

A. Cortés-Avizanda · J. A. Donázar
Department of Conservation Biology, Estación Biológica de Doñana, CSIC. Americo Vespucio s/n, 41092 Sevilla, Spain

J. A. Donázar
e-mail: donazar@ebd.csic.es

H. M. Pereira
German Centre for Integrative Biodiversity Research (iDiv) Halle-Jena-Leipzig, Deutscher Platz 5e, 04103 Leipzig, Germany

Institute of Biology, Martin Luther University Halle-Wittenberg, Am Kirchtor 1, 06108 Halle (Saale), Germany
e-mail: hpereira@idiv.de

H. M. Pereira, L. M. Navarro (eds.), *Rewilding European Landscapes,*
DOI 10.1007/978-3-319-12039-3_5, © The Author(s) 2015

however, alter the spatial-temporal nature of trophic resources with strong consequences at individual, population, community and ecosystem levels. The conservation of these charismatic species in rewilded European landscapes should rely on wild ungulate expansion, the recovery of large carnivore populations and, in more humanized areas, the promotion of traditional extensive agro-grazing systems limiting artificial feeding activities. In this way, it may be possible to combine both the historically recognized ecosystem services provided by vultures (elimination of undesirable remains, nutrient cycling) with new recreational services (conferring aesthetical value to the environment) while providing economic benefits to rural societies. Vultures and other scavengers, because they exploit space at a huge scale, are singular actors within a rewilded Europe. Their conservation, and that of the ecological processes in which they are involved, requires large-scale approaches surpassing those limits imposed by administrations, habitats and even biomes.

Keywords Carrion pulsed resource · Ecosystem services · Guilds · Predictability · Vulture restaurants · Wild ungulates

5.1 Introduction: Rewilding Ecosystem Services, Not Only Vertebrate Populations

Although the concept of "rewilding" is usually associated to restoring populations of symbolic species, it is obvious that it cannot be understood without the parallel amendment to the structure and functioning of ecosystems (see Chap. 1). In this sense, there are some ecological processes transverse to habitat structure, ecosystems and biomes, which are key to maintaining both complex food webs and the viability of populations of organisms. The large avian scavengers in Europe could serve as a paradigmatic case in this regard. Due to the early and rapid transformation of European landscapes, the decline of large scavenger populations occurred prior to that on other continents (Bijleveld 1974), thus breaking the ancient alliance between traditional agro-pastoral practices and the existent large populations of these birds. However, since the late 20th century to today, it is precisely in Europe that some of the most representative and healthiest populations of vultures and other Old World avian scavengers likely reside, in comparison to other locations where a massive population declines have occurred, such as in Africa and Asia. It is also in Europe where there are likely more active recovery programmes devoted to conserving scavenger' populations mainly based on reintroductions and renewed recognition of the ecosystem, cultural and economic services that scavengers may provide.

Thus, it may be deduced that the rewilding of Europe, specifically in the case of avian scavenger species, has already begun. This process is often accompanied by profound changes in the carrying capacity of the environment (e.g., through supplementary feeding programs). While undeniably profound changes are occurring that may make the maintenance of populations of birds of prey and the services they provide unstable, it is expected that in a few decades large areas of the European continent will abandon traditional grazing activities (see Chap. 1). On the other hand, other regions will continue the intensive occupation and use of land, which may impose potentially greater impacts on natural systems (e.g., agricultural intensification, growth of urban areas) (Deinet et al. 2013).

In this chapter, our goal is to explore, on the basis of existing information, how top scavengers fit into a wilder Europe. To do this, we first examine the role of availability of carrion resources in the maintenance of ecosystem functioning. Then, we examine the implications of the creation of supplementary feeding stations (so-called vulture restaurants). In addition, we will describe how the relationship between humans and vultures has evolved, identifying ecosystem services provided by these charismatic species from the past to the current date. Finally, we propose that the conservation of top scavenger species and the maintenance of natural complex ecological process linked to the exploitation of unpredictable carrion resources in rewilding the European landscapes should rely on favouring wild ungulates expansion, the recovery of predator populations and the promotion of traditional extensive grazing practices.

5.2   The Role of Carcasses Within Ecosystems

Carcasses are resources whose role is often minimized in ecological theory (DeVault 2003). Since animals can die at any place and time, a carcass is considered a "prize" for many species that obtain some benefit. From carnivores and avian scavengers to the microcosm, a large number of species, including plants, can benefit from the appearance of carcasses. Consumers can exhibit numeric and/or functional responses including behavioural changes, and increases in their reproductive rates in response to carcasses. A possible increase in the populations of consumers of carrion may trigger changes in the interactions between them and their prey, predators, competitors and parasites. In addition to these direct effects, indirect effects can occur that can elicit "cascading" effects through food chains, both bottom-up and top-down, ultimately affecting the community structure the ecosystem functioning (Cortés-Avizanda 2011; Yang et al. 2008; Fig. 5.1).

From an abiotic point of view, it is noteworthy that the appearance of a carcass represents a key natural "disturbance" in the composition of soils and plant communities, because it means a sudden availability of nutrients (Towne 2000; Melis et al. 2007). In fact, based on the high levels recorded of soil nutrients deposit, biomass production and activity of edaphic fauna, the experts have defined the carcasses as "islands of decomposition" (Carter et al. 2007; Selva and Cortés-Avizanda 2009; Fig. 5.1). This increase in soil nutrient concentration in the vicinity of a carcass shows a gradient decreasing towards the periphery and which may persist for several years (Towne 2000; Danell et al. 2002; Melis et al. 2007). For vegetation, the changes are more drastic and can occur both in terms of biomass and community structure. The "islands of decomposition" represent a resource of high quality and low competition, favouring the establishment of pioneer species. In general, it has

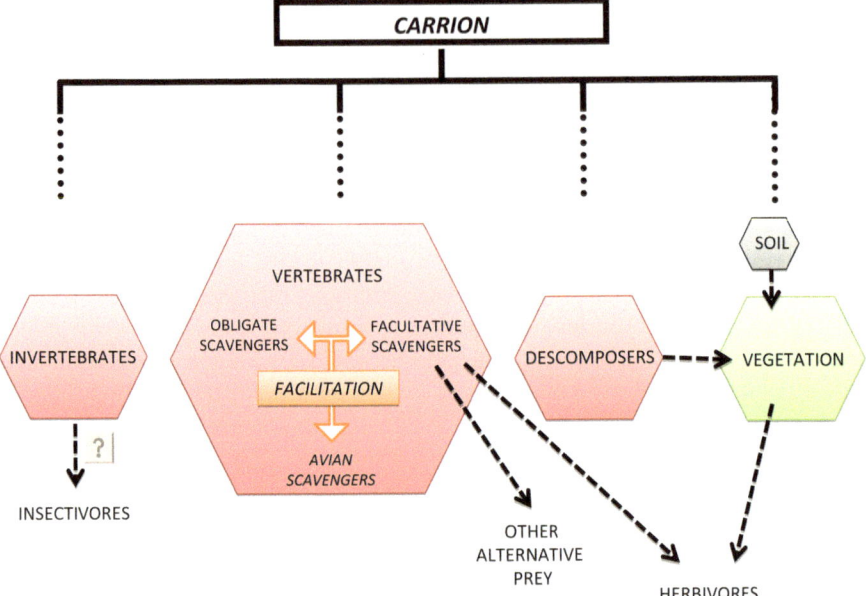

Fig. 5.1 Conceptual model of the energy flow and the ecological relationships that are established with the appearance of a carcass in the wild. The arrows indicate the well-known effects of some trophic groups on the others. Here it is also presented the example of the avian scavengers. The "?" indicate hypothetical effects that have not yet been studied in detail. (Based on Selva and Cortés-Avizanda 2009)

been well-described that one year after the occurrence of a carcass the richness, the diversity of species, the biomass and the vegetation cover can increase significantly in the vicinity of the carcass (Towne 2000). This effect depends on the region where it occurs and other numerous factors such as the climate, temperature and soil type. For example, if the carcass occurs in a homogeneous system or an unproductive one, then carcasses can represent a refuge for many plant species leading to an increase in the spatial diversity of the plant community (Towne 2000; Danell et al. 2002; Selva and Cortés-Avizanda 2009).

Relative to direct consumers, it has been reported that the diversity and complexity of microorganisms and invertebrates exploiting dead animals can acquire an extraordinary value (Sikes 1994). More than 500 species of arthropods (and of these, 422 species of insects) have been described in a carcass during different stages of decomposition (Payne 1965), varying by geographic region and environmental conditions (Amendt et al. 2004; Selva and Cortés-Avizanda 2009; Fig. 5.1). Many organisms show a relative specialization in relation to the type of carrion and its decomposition phase. Those organisms may respond to carcass appearance and further trophic cascades may elicit (Payne 1965; Amendt et al. 2004; and see examples in Selva and Cortés-Avizanda 2009). Among vertebrates, carcasses may be consumed by over 30 species of birds and mammals (Houston, 1979; DeVault et al. 2003; Selva 2004; Selva and Fortuna 2007). Most of these organisms are facultative scavengers, which are opportunistic carrion eaters, especially when their main prey become scarce. For instance, it is well-known that in temperate forests during harsh wintering conditions (low temperatures, thick snow cover and low availability of small mammals) predators significantly increase feeding on carcasses (Heinrich 1988; Jedrzejewski et al. 1993; Selva 2004; Selva et al. 2003, 2005; Cortés-Avizanda et al. 2009a).

Despite the importance of these many scavenger species, the avian scavengers are the organisms that have attracted the most attention because of their spectacular nature and their close relationship with humans as effective providers of ecosystem services (see below). These species have evolved different behavioural skills and morphology, allowing their coexistence through the sharing of trophic resources (Kruuk 1967; Root 1967; König 1983; Hertel 1994; Hertel and Lehman 1998; Blondel 2003; Cortés-Avizanda et al 2014; Fig. 5.2). The functioning of the guild is driven by both positive ecological processes (facilitation) and the most obvious competitive (negative) interspecific relationships (Donázar 1993; Cortés-Avizanda et al. 2012). Facilitatory processes have been proposed to follow two opposite paths within this guild: small-body-sized facultative scavengers landing earlier at carcasses can increase the chances of carcass detection by larger (specialist) vultures (local enhancement) whereas large vultures dismember the carcass thus allowing smaller scavengers to profit from the resource (trophic advantage; Kruuk 1967; König 1974, 1983; and see details in Cortés-Avizanda et al. 2012). As a consequence of this efficient process, carcasses of medium-sized animals are consumed in a very short time (Selva 2004; Cortés-Avizanda et al. 2012; Cortés-Avizanda et al. 2014).

5.3   Vultures and Humans: An Unstable Alliance

Top scavengers ("true" vultures, Accipitridae) have been evolutionarily dependent on carcasses of large animals, mainly ungulates, grazing in open areas of southern Europe, Africa and central and southern Asia (Houston 1974, 1979; Donázar 1993). They have in common extreme adaptations and skills aimed to locate scarce and unpredictable sources of food (Houston 1979; Cortés-Avizanda et al. 2014). However, because of the rampant humanization of ecosystems, this natural scenario no longer exists, apart from some strictly protected areas, especially in African countries where large herds of ungulates subsist (Cortés-Avizanda et al. 2011; Fig. 5.2). Thus, following the progressive eradication of most native populations of wild ungulates (e.g. Chap. 8), guilds of avian scavengers have become largely dependent on livestock carcasses associated with human activities (Mundy et al. 1992; Donázar 1993). This scenario has likely remained almost unchanged since the so-called "Neolitic revolution" (i.e., a process in which agrarian societies began to substitute hunting around 8500 BC with agricultural practices including the domestication of the herbivores) in many regions of the Old World, and certainly in southern Europe, where the agro-grazing traditional economies remained unchanged (Donázar et al. 1996a, 2009; Olea and Mateo-Tomas 2009; Fig. 5.3).

Fig. 5.2 a Egyptian vulture (*Neophron percnopterus*) and **b** Black kite (*Milvus migrans*) in the Mauritanian Sahel. In Africa the availability of trophic resources is not limiting and therefore scavengers show a homogeneous spatial distribution. (Cortés-Avizanda et al. 2011; Photo credit: Jose Ramón Benitez)

After the industrial revolution this scenario changed dramatically. During the 19th and the first half of the 20th century, a utilitarian view of nature became common. "Harmful" species were more efficiently prosecuted and raptor populations, including scavengers, declined steeply across Europe and in many regions to full extinction (Bijleveld 1974). In parallel, a key resource for carrion-eaters, the wild ungulates, were

Fig. 5.3 Avian scavengers and traditional extensive grazing. After the eradication of native populations of wild ungulates, avian scavengers have become dependent of traditional livestock carcasses. (Photo credit: Iosu Anton)

also decimated by hunting and poaching (the main decline of ungulate populations in Europe took place during the last centuries after the generalized use of firearms), disappearing from most of the European landscapes (see e.g. Chapman and Buck 1910; Deinet et al. 2013 and references therein). Carcasses of domestic ungulates were probably still plentiful until well into the 20th century but the transformation of farming to intensive practices and the abandonment of extensive grazing reduced their availability in the last decades (Donázar et al. 1996a). This dramatic scenario began to improve from the 1960s onwards. Given the high public profile of many large avian scavengers, local administrations and conservationist groups created a number of feeding stations (also known as "vulture restaurants"; Fig. 5.4) to supply food and help re-establish the decimated populations of these species (Bijleveld 1974 and see below; Fig. 5.4). But perhaps much more importantly, legal protection curbed non-natural mortality allowing scavenger populations to quickly recover, mainly in the Iberian Peninsula and Southern France (Donázar and Fernández 1990; Donázar et al. 1996a, b; Slotta-Bachmayr et al. 2004). The griffon vulture (*Gyps fulvus*) was the most favoured species, its populations having dramatically increased in numbers in the Iberian Peninsula and France (BirdLife International 2004). Populations of Cinereous vulture (*Aegypius monachus*) and Bearded vulture (*Gypaetus barbatus*) have also increased in numbers, albeit more moderately, in some European regions (Margalida and Heredia 2005; Moreno-Opo 2007; Dobado et al. 2012). By contrast, the numbers of small-sized scavengers such as Egyptian vultures (*Neophron percnopterus*) and Red (*Milvus milvus*) and Black (*Milvus migrans*) kites continue to decline (Viñuela et al. 1999; Del Moral and Martí 2001, 2002).

Fig. 5.4 Supplementary feeding stations devoted for vultures. The supplementary feeding stations are a worldwide conservation tool to recover endangered populations of avian scavengers. Usually they are fenced sites where local farmers and rangers dispose the carcasses. **a** A vulture restaurant devoted to Egyptian vulture in Fuerteventura (Canary Island) where the dominant griffons are not presented. **b** A group of Egyptian vulture feeding at a vulture restaurant also in Fuerteventura. (Photo credit : a) José Antonio Donázar and b) Manuel de la Riva)

A decade ago the appearance of the Bovine Spongiform Encephalopatie (BSE) changed the picture abruptly. New sanitary regulations driven by the European Union banned the abandonment of livestock carcasses in the field, and consequently, the availability of food resources declined in some regions by more than 80 %. The consequences for vulture populations are still being evaluated, but spatial distribution, breeding success and survival seem to have been affected. This is particularly noticeable in those species with greater dependence on carcasses of large animals such as griffon vultures (see reviews in Donázar et al. 2009a). To complicate things even more, after these sanitary measures a conflict between farmers and vultures developed in Mediterranean regions (particularly Spain and southern France). Griffon vultures are known to occasionally kill and consume diminished livestock but the number of cases reported increased sharply after the implementation of sanitary regulations (Margalida et al. 2014). No doubt that this trend was largely the result of a social contagion driven by misinformed media, but regardless, a true conflict arose compromising decades of conservation measures aimed to restore avian scavenger populations. The consequence was that hundreds of individuals of griffon and other species of vultures and facultative scavengers perished by poison, while public opinion called for a quick fix consisting of artificially feeding vultures to distract their attention from live prey (Margalida et al. 2010).

After these events and due to the widespread pressures coming from researchers, policy conservation managers and farmers, recent European regulations have increasingly allowed the disposal of livestock carcasses for consumption by vultures (Donázar et al. 2009a, b; Margalida et al. 2010). The adoption of this new legislation takes time, especially for the transposition of EU laws to individual country governments and then in many cases from the country level to local government departments. In general, the legal framework is cumbersome such that administrations are faced with complex and time-intensive processes. Consequently, the European regional governments (e.g. in Spain), have adopted a common strategy whereby just a few widespread sites -in theory under strict veterinary control- are supplied with carcasses for scavengers. Thus, vultures have access to restaurant networks with large amounts of food (Cortés-Avizanda et al. 2010, 2012). The effects of this conservation strategy, which is not new and has now become widespread, may be substantial from both population and ecological points of view.

5.4 Vulture Restaurants and the Loss of a Pulsed Resource

Supplementary feeding stations or vulture restaurants have been considered for decades to be a key management tool for the conservation of scavenger bird populations (Bijleveld 1974; Houston 1987; Piper 2006). Species managed by supplementary feeding include: the California Condor (*Gymnogyps californianus*) in Western North America; the King Vulture (*Sacorramphus papa*) in Belize; and the Cape Vulture (*Gyps coprotheres*) and bearded vulture in Southern Africa (Wilbur and Jackson 1983; Houston 1987; Brown 1990; Mundy et al. 1992). In Eurasia, vulture restaurants have mainly targeted populations of bearded,

Egyptian, griffon, and cinereous vultures (Donázar 1993; Donázar et al. 2009a, 2010; Cortés-Avizanda et al. 2010; Fig. 5.4). Supplementary feeding stations have also been considered as key tools in the recovery of the critically endangered Asian vulture populations after the catastrophic declines caused by the veterinary treatment of livestock with diclofenac (i.e., NSAID, a non-steroidal anti-inflammatory drug, see Gilbert et al. 2007; Prakash et al. 2012 and references therein). In general, it is assumed that this management tool provides a number of benefits for the conservation of the target species: an increase of food availability, reduction of the risk of poisoning and persecution, and an increase in the availability of micronutrients (calcium) (Piper 2006).

Supplementary feeding stations, however, change the spatio-temporal distribution of a food resource that is otherwise unpredictable and ephemeral. As we have described above, in natural systems, carcasses resemble other trophic pulsed resources such as tree-masting or insect explosions (Ostfeld and Keesing 2000; Rose and Polis 1998). Currently, however, carcasses have become predictable and clumped via this human intervention and the consequences of this deep alteration of habitat quality have been largely ignored, under the assumption that only positive effects would occur. We are now increasingly aware that these positive effects exist but that there are also other negative effects that may override the positive conservation effects on the target species. For instance, it has been demonstrated that proximity to vulture restaurants promotes communal roosts and long-term territory maintenance in the Egyptian vulture increasing the probability of frequent visits by breeding adults from nearby territories (Grande et al. 2009; Benítez et al. 2009; García-Heras et al. 2013; Lopez-Lopez 2014). In the case of the Pyrenean bearded vultures, supplementary feeding stations have led to high rates of immature survival thus dampening the effects of indirect persecution (poisoning) and increasing population viability (Oro et al. 2008). Conversely, negative population effects are also apparent. Negative density-dependent decreases of productivity and appearance of unusual breeding units (polyandrous trios) have been detected in the vicinity of bearded vulture feeding places (Carrete et al. 2006a, b). Unwanted effects have also been detected in the structure and functioning of guilds. Vulture restaurants favour the gathering of individuals of the dominant species (e.g., the griffons), which monopolize food to the detriment of small and less competitive and often more endangered scavenger species (Cortés-Avizanda et al. 2010; 2012). The predictable nature (in space and time) of carrion disposed in these places disrupts ecological processes provoking a reduction of guild diversity and the loss of intraguild facilitatory processes because dominant specialist species (griffons) arrive early and in larger numbers (Cortés-Avizanda et al. 2012). Less competitive species may congregate at feeding places but they may be unsuccessful at feeding, whereby the feeding place would become an ecological trap (Cortés-Avizanda et al. 2012 and authors' unpublished data; Fig. 5.5).

Finally, the aggregation of food resources at scavenger feeding places may have consequences on non-scavenging species, permeating to other ecological levels. Facultative species consume carcasses less efficiently and more slowly than the specialists or strict carrion-eaters and most importantly, they do not feed only on carcasses but also rely on small prey. Therefore, predation pressure can increase on passerine species breeding in the vicinities of vulture restaurants and carcass

Fig. 5.5 Randomly distributed vs. predictable resources. Unpredictable trophic resources allow the occurrence of facilitatory processes promoting the bio-diversity and the coexistence of species within an Old World avian scavenger guild: E.g. **a** A Bearded vulture (*Gypaetus barbatus*) sharing a carcass with a common raven (*Corvus corax*) and **b** A griffon vulture (*Gyps fulvus*) sharing a carcass with two Egyptian vultures (*Neophron percnopterus*). **c** Group of griffon vultures at supplementary feeding stations. Large amount of carrion clumped at these predictable sites favours the aggregations of hundreds of individuals (Photo Credit: Jordi Bas **a** and Antonio Atienza **b** *and* **c**)

accumulations can change the spatial distribution of herbivore mammals (Cortés-Avizanda et al. 2009a, b; Wilmers et al. 2003). This phenomenon may be pronounced at high latitudes, where cold temperatures during long winters slow the activity of microorganisms and invertebrates and where there is a lack of specialist carrion-eaters that would otherwise quickly deplete the carrion (Selva and Cortés-Avizanda 2009; Cortés-Avizanda 2011; Cortés-Avizanda et al. 2009a).

5.5   How do Vultures Fit into a Rewilding Continent?

According to Navarro and Pereira (see Chap. 1), the decline in the number of extensive livestock in Europe was 25 % between 1990 and 2010. From this assessment
two interpretations are possible: on the one hand, the abandonment of traditional
grazing involves a significant reduction in livestock and domestic food sources for
vultures, but on the other hand, the landscape abandonment may contribute positively to the expansion of wild ungulate populations in many rural areas of Europe,
notably in mountain ranges. Blázquez-Alvarez and Sánchez-Zapata (2009) showed
that the number of wild ungulates hunted in Spain went from 60,000 in around
1980 to 200,000 in 2005. The current area of distribution occupied by wild ungulate species has also spread (e.g., 70–75 % of the Spanish territory, Sánchez-Zapata
et al. 2010) such that up to six different species of ungulates can be found in some
mountain areas. Similar trends can be found in other Mediterranean and temperate
regions of Europe (Milner et al. 2006).

   Increasingly larger populations of wild ungulates are allowing a return to natural diets of specialists and facultative scavenger species (see Moreno-Opo et al.
2007; Sánchez-Zapata et al. 2010). The availability of wild ungulate carcasses for
vultures would increase even more if future rewilding processes lead to the expansion of large carnivores which would contribute a regular supply of random carcasses (Selva 2004; Blázquez-Alvarez and Sánchez-Zapata 2009). More important
than the global availability of carcasses may be the ways in which the existence of
large carnivores can change the temporal and spatial distribution of the resource
and the associated consequences. It has been argued that predation can add stability to trophic networks buffering those oscillations linked to temporally-pulsed
events of carrion availability such as those determined by climatic events, diseases
and hunter-kills (Wilmers and Getz 2005; Wilmers and Post 2006). Currently, and
because of persecution, wolves in European landscapes are restricted to forests and
mountains, and therefore most of the species benefiting from predated animals are
forest-living facultative scavengers (Selva 2004; Selva et al. 2003; 2005). It is expected, however, that in a more relaxed scenario carcasses of killed ungulates would
also be available in open biomes, thus being available to large avian scavengers,
and restoring the stability and the insurance of food chains and ecological processes
(Wilmers and Getz 2005; Tylianakis et al. 2010).

   A key question arises from a scenario of substitution of domestic by wild ungulates: can vulture populations survive in a future wilder scenario where most of the
carcasses are provided by wild herbivores? Recent research performed in North-
Eastern Spain by Margalida et al. (2011) based on the application of bio-inspired
computational models revealed that this may be the case for high mountain areas
where large and diverse populations of wild ungulates subsist. On the contrary, in
low-altitude areas with higher humanization densities, wild herbivores are scarce
and their populations would be insufficient to guarantee the long-term maintenance
of avian scavenger populations such as those of griffon, Egyptian and bearded
vultures. Therefore, these authors also suggest that in lowland areas of European
Mediterranean regions, carrion-eaters will still be dependent on resources provided

Fig. 5.6 Cultural services provided by birds benefit local communities. The delight with the beauty of vultures alone or soaring in large group at their breeding areas and/or feeding at vulture restaurants confer a value to environment, and attract birdwatchers from all over the world translating this cultural service into significant incomes via "ecoturism". (Photo credit: Jordi Bas)

by traditional extensive livestock and/or those supplies offered by conservation managers in vulture restaurants. From these results arise new questions for future research about how extensively rewilding should be facilitated, especially in those low-altitude humanized areas, in order to decrease vulture dependence on networks of large and fixed supplementary feeding stations.

The consumption of wild ungulate carcasses derived from hunting is not exempt of risk for avian scavengers. Birds ingesting pieces of hunting ammunition are exposed to lead intoxication (see review in Fisher et al. 2006; Mateo et al. 2007) to the extent that its incidence can be a serious threat to large-scale reintroduction projects such as that of the California condor in North America (Finkenlstein et al. (2012 and references therein). Lead levels are also very high in Europe, for instance Iberian griffon vultures show seasonal and spatial variations according to the rate of consumption of wild ungulates (García-Fernández et al. 2005). Moreover, carcasses resulting from hunting activities frequently accumulate at a few points at the end of hunting activities (Wilmers et al. 2003), which mimics the predictability offered by vulture restaurants and may result in similar negative effects (see above). Under a future scenario of a wilder Europe, we may promote two lines of action to curb the poisoning of scavengers: the consideration of a wilder Europe with zones limiting or lacking in hunting activities and in parallel the encouragement of the traditional extensive grazing currently under decline.

Fig. 5.7 Hotspots of abandonment and rewilding and distribution of the most important breeding areas for top scavenger: Giffon vultures and the endangered Egyptian, bearded and cinereous vultures. The map in the centre, relative to rewilding, shows areas categorized as "agriculture" in 2000 that are projected to become rewilded or afforested in 2030 with the CLUE model (see Chap. 1 for mapping method). The maps showing the scavenger distributions are based on information available in the Spanish Atlas of Breeding Birds (Del Moral and Martí 2004). (Species drawings: Juan Varela)

5.6 New Services Provided by Vultures

Although there is a strong competition "between" and "within" guilds (bacteria, invertebrates, carnivores, birds) for pulsed carrion resources (see reviews of Root 1967; Jaksic 1981; Schluter and Ricklefs 1993; Blondel 2003), this does not diminish the fact that potentially harmful species (like rats and feral dogs) prosper with local food abundance (Markandya et al. 2008), carrying potentially infectious diseases (Blount et al. 2003). Therefore, a historical scavengers/humans relationship was built, resulting in a sort of mutualism: humans provide trophic resources and vultures eliminate undesirable remains and control the concentration of disease (Deygout et al. 2009; Margalida et al. 2011; Cortés-Avizanda et al. 2012). Currently, and because both sanitary regulations impose the efficient elimination of livestock remains, and modern societies require that industrial activities are performed in a way that minimizes their ecological footprint, these services have acquired a new dimension: vultures can remove carcasses at zero cost whereas the elimination of livestock carcasses by means of industrial procedures entails high expenditures (transport: 20 €/animal; destruction: 96 €/t; see Donázar et al. 2009a, b). Moreover, the CO_2 emissions derived from the transport and burning of carcasses are not negligible. Consequently, modern farming economies may still find vultures to be useful allies in sustaining traditional uses of Mediterranean landscapes (Deygout et al. 2009).

In modern societies, vultures, like other large birds, are highly attractive wildlife, providing several recreational services: the delight experienced from the beauty of a vulture in flight or soaring in large groups, or the observation of feeding behaviour and interactions, among other, bring great value to the environment (Fig. 5.6). These services may be translated into significant incomes derived from "ecotourism" (see Chap. 3). For instance, it is estimated that visitors for griffon vulture watching in Israel provides around US$ 1.1–1.2 million per year (Becker et al. 2005) and the park "La falaise aux vautours" in the French Pyrenees receives 15,000–20,000 visitors per year whose primary motivation is wildlife viewing activities. Overall, the observation of scavenger breeding areas and vulture restaurants is an increasingly common activity within specialized touristic tours. This can improve the income of those southern European rural societies subject to profound environmental and socioeconomic changes.

To summarize, we can conclude that scavenging directly contributes to human well-being. Avian scavengers provide regulation/maintenance and cultural services due to their valuable role in the decomposition of carcasses, pest control, biodiversity maintenance, and tourism attraction (Haines-Young and Potschin 2013; Maes et al. 2013)

5.7   Discussion and Conclusion

Land abandonment is widespread in many European areas with poor agricultural development. In the face of this situation, the rewilding of abandoned landscapes (defined as the passive management of ecological succession, see Chap. 1; Fig. 5.7) is viewed as an opportunity to recover native biodiversity and ecological processes and provide a range of ecosystem services (Cramer et al. 2008; and see Chaps. 1, 3). In contrast, active management is aimed at the maintenance of low-intensity agriculture in order to conserve specific organisms and particular habitats, often working against succession processes (e.g. Pain and Pienkowski 1997). This dichotomy is subject to a heated debate with clear implications for European agricultural policies (see details in Merckx and Pereira in press). Within this scenario, the conservation of top scavengers, not only as charismatic species but also as key actors in complex ecological processes, imposes particular challenges. Vultures, as well as a panoply of facultative predatory-scavenger vertebrates, arthropods, and microorganisms, depend largely on the existence of ungulate and medium-size vertebrate carcasses, a resource that in natural conditions is unpredictable in space and time (i.e., by chance, see above). The importance of ecological processes linked to carcass decomposition and consumption has been historically neglected (DeVault 2003). Instead, carcasses have been considered undesirable and insanitary residuals whose common destination has been industrial destruction. Consequently, a key resource in European ecosystem functioning was sent to the incinerator following sanitary regulations dictated by an exaggerated precautionary principle (Donázar 2009b). Rewilding may now favour the increase of larger populations of wild ungulates as well as the expansion of large carnivores (see Chap. 4), which would contribute to the random nature of carcasses and uphold ecosystem functioning and community structures.

Other distinctive aspects of top scavengers, and vultures in particular, is that these species exploit landscapes at a scale much larger than humans do. Vultures' home ranges and foraging displacements may cover as many as hundreds of thousands of hectares. For instance, modern monitoring techniques based on GPS and satellite tracking have demonstrated continental-scale movements of individual birds (e.g. bearded vultures in Margalida et al. 2013; Egyptian vultures in Carrete et al. 2012; among others). This is the result of adaptations to searching for unpredictable resources (Hertel 1994; Cortés-Avizanda et al. 2014), originally ungulate herds moving across changing landscapes (Houston 1974), something that still occurs in modern Europe with transhumant livestock husbandries (i.e., movement of livestock between winter and summer pastures, Olea and Mateo 2009), but on a very limited basis today. Thus, vultures have space requirements that can scarcely be reduced to the small-scale of administrative limits prevailing in European landscapes. In this context the question that arises is how can we manage to fit vultures into the projected rewilded Europe? As mentioned above, recent studies related to vulture populations in the Spanish Pyrenees show that carcasses of wild ungulates are able to maintain vulture populations in the long-term (Margalida et al. 2011 and see above for details). However, the expansion of wild herbivores into the abandoned lands of many European mountains is already occurring and likely to increase in the future (see Chap. 8). However, to date, according to Margalida et al. (2011), although avian scavengers make large-scale seasonal movements from mountains to lowlands the recovery of wild ungulates in mountain areas would be insufficient to support the populations. At best, it is evident that the long-term viability of vulture populations restrained to mountain areas would be affected by constraints derived from small population size (Margalida et al. 2011; Donázar et al. 2009a). On the other hand, many healthy vulture populations thrive in very diverse lowland agro-grazing systems where most of the carcasses are provided by extensive livestock and wild medium-sized prey (notably wild rabbits *Oryctolagus cuniculus*) (Donázar et al. 1996b; Moreno-Opo et al. 2010; Carrete et al. 2007). Therefore, under the scenario of a future wilder Europe and in conjunction with the expansion of wild ungulates and predators, the encouragement of traditional extensive grazing should be prioritised, especially in those lowland humanized areas of agro-grazing systems where arrival of wild ungulates may be no possible.

Overall, the conservation of vultures and other avian scavengers requires a necessary equilibrium between the recovery of wild ungulates in remote (and rewilded) areas, the existence of traditional agro-grazing systems in lowland regions and, the modifications/adaptations of sanitary laws in order to allow the abandonment of livestock carcasses freely in the wild (even outside of protected and defined areas). Recent European regulations are opening the way for this approach and within this context, the maintenance of supplementary feeding stations (vulture restaurants) is not a desirable conservation strategy because of the negative population, community and ecosystem effects that appear to surpass the positive effects linked to the mere improvement of demographic parameters (see above for references). Therefore, this scenario represents a great opportunity to reclaim future conservation strategies considering community and ecosystem perspectives. New policies should focus on: (1) the conservation of foraging and breeding behaviour of hundreds of scavenger as well as the ecosystem restoring by promoting the availability of natural random carcasses; (2) highlighting the importance of the regulating services derived from those feeding behaviours such as the reduction of transmission of animal diseases; (3) the new ecosystem services not related directly to species conservation but that may imply greater incomes to local economies from ecotourism and associated with recreational activities in natural areas; and (4) the consideration of non-economic value that these species may provide such as the existence values and use them to educate society on conservation.

The scenario described thus far applies to those regions of Western Europe (the Iberian Peninsula, Southern France and other Mediterranean regions) where large scavenger populations have persisted (Donázar et al. 2009a). A totally different picture exists in central and Eastern Europe, where the populations of vultures and other carrion-eaters were virtually extirpated during the course of the twentieth century. There, passive rewilding of large regions probably will not lead to short or medium-term recovery of top scavenger/vulture populations and the associated scavenging processes because of the extremely low rate of spatial expansion of populations of these long-lived organisms that is explained by extreme natal philopatry. In this case, active measures like reintroduction are required and of course, the elimination of those limiting factors, mainly direct and indirect persecution, that once determined the populations' demise. Rewilding also considers the reintroduction of species and initial support with supplementary feeding stations. In this case, we suggest that those programs be based on rigorous scientific population viability studies and that food supplies, if necessary, are provided under careful adaptive management for greater effectiveness of rewilding conservation decisions (Cortés-Avizanda et al. 2010; McCarthy and Possingham 2007; Possingham et al. 2001)

Finally, while many mammal species and other vertebrates such as large ungulates and carnivores, may be viable in large mountain areas subjected to passive management (Fig. 5.7; and see details in Chap. 1), populations of flying organisms (covering large distances on a daily basis) would nevertheless require broader approaches dealing with both conservation aims and common policies in rewilded regions and in areas where traditional agro-grazing activities are maintained. Moreover, it must not be forgotten that top scavenger breeding in remote cliffs are providing key ecological services not only in high-mountain pasturelands but also in lowland grazing areas. These reflections suggest that new questions may arise along with new challenges, especially those related to how the maintenance of ecologically functional populations of large body-sized and long-lived organisms fits within the current rewilding concept (see also Chaps. 4 and 8). To attend to these targets and requirements effectively is to ensure the persistence of the alliance between humans and vultures that has allowed the survival of these charismatic birds for millennia.

Acknowledgments We wish to thank Jose Antonio Sánchez Zapata, Roger Jovani, and Nuria Selva who kindly discussed with us some of the ideas presented here and provided helpful comments on the earlier versions of the chapter. ACA was funded by a PostDoc grant from the Fundação para a Ciência e a Tecnologia (FCT) (SFRH/BPD/91609/2012), and JAD by the project CGL2012-40013-C02-01. Isa Afan from the Laboratorio de SIG y Teledetección, Estación Biológica de Doñana, CSIC (LAST-EBD), built the maps of species distribution. We also thank the photographers, Jose Ramón Benítez, Iosu Anton, Jordi Bas, and Antonio Atienza, Manuel de la Riva for their contribution to this chapter.

References

Amendt, J., Krettek, R., & Zehner, R. (2004). Forensic entomology. *Die Naturwissenschaften, 91,* 51–65.

Becker, N., Inbar, M., Bahat, O., Choresh, Y., Ben-Noon, G., & Yaffe, O. (2005). Estimating the economic value of viewing griffon vultures *Gyps fulvus:* A travel cost model study at Gamla Nature Reserve, Israel. *Oryx, 39,* 429–434.

Benítez, J. R., Cortés-Avizanda, A., Ávila, E., & García, R. (2009). Effects of the creation of a supplementary feeding station for the conservation of Egypttian vulture (*Neophron percnopterus*) population in Andalucia (southern Spain), in: Donázar, J.A., Margalida, A. & Campión, D. (Eds.),Vultures, feeding stations and sanitary legislation: a conflict and its consequences from the perspective of conservation biology. Sociedad de Ciencias Aranzadi San Sebastián, Spain, pp. 276–291.

Bijleveld, M. (1974). *Birds of prey in Europe*. London: Macmillan Press.

BirdLife International (2004). *Birds in the European Union: A status assessment.* BirdLife International. Wageningen, The Netherlands.

Blázquez-Alvarez, M., & Sánchez-Zapata, J. A. (2009). The role of wild ungulates as a resource for the community of vertebrate scavengers. In J. A. Donázar, A. Margalida, & D. Campión (Eds.), *Vultures, feeding stations and sanitary legislation: A conflict and its consequences from the perspective of conservation biology* (pp. 308–327). Munibe 29 (suppl.), Sociedad de Ciencias Aranzadi San Sebastián, Spain.

Blondel, J. (2003). Guilds or functional groups: Does it matter? *Oikos, 100,* 223–231.

Blount, J. D., Houston, D. C., Moller, A. P., & Wright, J. (2003). Do individual branches of immune defence correlate? A comparative case study of scavenging and non-scavenging birds. *Oikos, 102,* 340–350.

Brown, C. J. (1990). An evaluation of supplementary feeding for Bearded Vultures and other avian scavengers in the Natal Drakensberg. *Lammergeier, 41,* 30–36.

Carrete, M., Donázar, J. A., & Margalida, A. (2006a). Density-dependent productivity depression in Pyrenean Bearded Vultures: Implications for conservation plans. *Ecological Applications, 16,* 1674–1682.

Carrete, M., Donázar, J. A., Margalida, A., & Bertran, J. (2006b). Linking ecology, behaviour and conservation: Does habitat saturation change the mating system of bearded vultures? *Biology Letters, 2,* 624–627.

Carrete, M., Grande, J. M., Tella, J. L., Sánchez-Zapata, J. A., Donázar, J. A., Díaz-Delgado, R., & Romo, A. (2007). Habitat, human pressure, and social behaviour: Partialling out factors affecting territory extinction in the Egyptian vulture. *Biological Conservation, 136,* 143–154.

Carrete, M., Bortolotti, G. R., Sánchez-Zapata, J. A., Delgado, A., Cortés-Avizanda, A., Grande, J. M., & Donázar, J. A. (2012). Stressful conditions experienced by endangered Egyptian vultures on African wintering areas. *Animal Conservation, 16,* 353–358.

Carter, D. O., Yellowlees, D., & Tibbett, M. (2007). Cadaver decomposition in terrestrial ecosystems. *Die Naturwissenschaften, 94,* 12–24.

Chapman, A., & Buck, G. (1910). *Unexplored Spain.* London: Arnold.

Cortés-Avizanda, A., Selva, N., Carrete, M., & Donázar, J. A. (2009a). Effects of carrion resources on herbivore spatial distribution are mediated by facultative scavengers. *Basic and Applied Ecology, 10,* 265–272.

Cortés-Avizanda, A., Carrete, M., Serrano, D., & Donázar, J. A. (2009b). Carcasses increase the probability of predation of ground nesting birds: A caveat regarding the conservation value of vulture restaurants. *Animal Conservation, 12,* 85–88.

Cortés-Avizanda, A., Donázar, J. A., & Carrete, M. (2010). Managing supplementary feeding for avian scavengers: Guidelines for optimal design using ecological criteria. *Biological Conservation, 143,* 1707–1715.

Cortés-Avizanda, A. (2011). The ecological and conservation effects of trophic resource predictability: carcasses and vertebrate communities. PhD Thesis. Universidad Autónoma Madrid, Spain.

Cortés-Avizanda, A., Jovani, R., Carrete, M., & Donázar, J. A. (2012). Resource unpredictability promotes species diversity and coexistence in an avian scavenger guild: A field experiment. *Ecology, 93,* 2570–2579.

Cortés-Avizanda, A., Jovani, R., Donázar, J. A. & Grimm, V. (2014). Birds sky networks: How do avian scavengers search for carrion resource. *Ecology, 95,* 1799–1808.

Cramer, V. A., Hobbs, R. J., & Standish, R. J. (2008). What's new about old fields? Land abandonment and ecosystem assembly. *Trends in Ecology & Evolution, 23,* 104–112.

Danell, K., Berteaux, D., & Braathen, K. A. (2002). Effect of muskox carcasses on nitrogen concentration in tundra vegetation. *Arctic, 55,* 389–392.

Deinet, S., Ieronymidou, C., McRae, L., Burfield, I. J., Foppen, R. P., Collen, B., & Böhm, M. (2013). *Wildlife comeback in Europe: The recovery of selected mammal and bird species. Final report to Rewilding Europe by ZSL, BirdLife International and the European Bird Census Council.* London: ZSL.

Del Moral, J. C., & Martí, R., (2001). *El Buitre leonado en la Península Ibérica. Monografía n°7.* SEO/BidLife, Madrid.

Del Moral, J. C., & Martí, R. (2002). *El Alimoche en España y Portugal. Monografía n°8.* SEO/BidLife, Madrid.

Del Moral, J. C., & Martí, R. (2004). Atlas de aves reproductoras de España. Dirección General de Conservación de la Naturaleza-SEO. Madrid, Spain.

DeVault, T. L., Rhodes, O. E., & Shivik, J. A. (2003). Scavenging by vertebrates: Behavioral, ecological, andevolutionary perspectives on an important energy transferpathway in terrestrial ecosystems. *Oikos, 102,* 225–234.

Deygout, C., Gault, A., Sarrazin, F., & Bessa-Gomes, C. (2009). Modeling the impact of feeding stations on vulture scavenging service efficiency. *Ecological Modelling, 220,* 1826–1835.

Dobado, P. M., Díaz, F. J., Díaz-Portero, M. A., García, L., Luque, E., Martín, J., Martínez, P., & Arenas, R. M. (2012). El buitre negro Aegypius monachus en Andalucía (España). In P. M. Dobado & R. M. Arenas (Eds.), *The Black Vulture: Status, conservation and studies* (pp. 13–37). Consejería de Medio Ambiente de la Junta de Andalucía, Sevilla, Spain.

Donázar, J. A. (1993). Los buitres ibéricos. In J. M. Reyero (Ed.), *Biología y Conservación.* Madrid, Spain.

Donázar, J. A., & Fernández, C. (1990). Population trends of Griffon Vultures (*Gyps fulvus*) in northern Spain between 1969 and 1989 in relation to conservation measures. *Biological Conservation, 53,* 83–91.

Donázar, J. A., Naveso, M. A., Tella, J. L., & Campión, D. (1996a) Extensive grazing and raptors in Spain. In D. J. Pain & M. W. Pienkowski (Eds.), *Farming and birds in Europe* (pp. 117–149). Cambridge: Academic.

Donázar, J. A., Ceballos, O., & Tella, J. L. (1996b). Communal roost of Egyptian vultures (*Neophron percnopterus*): dynamics and implications for the species conservation. In J. Muntaner, & J. Mayol (Eds.), *Biology and conservation of mediterranean raptors (1994)* (pp. 189–202). SEO/BirdLife, Monography 4, Madrid.

Donázar, J. A., Margalida, A., & Campión, D. (2009a). *Vultures, feeding stations and sanitary legislation: A conflict and its consequences from the perspective of conservation biology.* Sociedad de Ciencias Aranzadi San Sebastián, Spain.

Donázar, J. A., Margalida, A., Carrete, M., & Sánchez-Zapata, J. A. (2009b). Too sanitary for vultures. *Science, 326,* 664.

Donázar, J. A., Cortés-Avizanda, A., & Carrete. M. (2010). Dietary shifts in two vultures after the demise of supplementary feeding stations: Consequences of the EU sanitary legislation. *European Journal of Wildlife Research, 56,* 613–621.

Finkelstein, M. E., Doak, D. F., George, D., Burnett, J., Brandt, J., Church, M., Grantham, J., & Smith, D. R. (2012). Lead poisoning and the deceptive recovery of the critically endangered California Condor. *Proceedings of the National Academy of Sciences of the United States of America, 109,* 11449–11454.

Fisher, I. J., Pain, D. J., & Thomas, V. G. (2006). A review of lead poisoning from ammunition sources in terrestrial birds. *Biological Conservation, 131,* 421–432.

García-Fernández, A. J., Martinez-Lopez, E., Romero, D., Maria-Mojica, P., Godino, A., & Jimenez, P. (2005). High levels of blood lead in griffon vultures (*Gyps fulvus*) from Cazorla Natural Park (southern Spain). *Environmental Toxicology, 20,* 459–463.

García-Heras, S., Cortés-Avizanda, A., & Donázar, J. A. (2013). Who are we feeding? Asymmetric individual use of surplus food resources in an insular population of the endangered Egyptian vulture *Neophron percnopterus*. *PLoS ONE, 8,* e80523.

Gilbert, M., Watson, R. T., Ahmed, S., Asim, M., & Johnson, J. A. (2007). Vulture restaurants and their role in reducing diclofenac exposure in Asian vultures. *Bird Conservation International, 17,* 63–77.

Grande, J. M., Serrano, D., Tavecchia, G., Carrete, M., Ceballos, O., Díaz-Delgado, R., Tella, J. L., & Donázar, J. A. (2009). Survival in a long-lived territorial migrant: Effects of life-history traits and ecological conditions in wintering and breeding areas. *Oikos, 118,* 580–590.

Haines-Young, R., & Potschin, M. (2013).Common International Classification of Ecosystem Services (CICES): Consultation on Version 4, August-December 2012. EEA Framework Contract No EEA/IEA/09/003.

Heinrich, B. (1988). Winter foraging at carcasses by three sympatric corvids, with emphasis on recruitment by the raven, (*Corvus corax*). *Behavioral Ecology and Sociobiology, 23,* 141–156.

Hertel, F. (1994). Diversity in body size and feeding morphology within past and present vulture assemblages. *Ecology, 75,* 1074–1084.

Hertel, F., & Lehman, N. (1998). A randomized nearest neighbor approach for assessment of character displacement: The vulture guild as a model. *Journal of Theoretical Biology, 190,* 51–61.

Houston, D. C. (1974). Food searching in griffon vultures. *East African Wildlife, 12,* 63–77.

Houston, D. C. (1979). The adaptations of scavengers. In A. R. E. Sinclair & M. N. Griffiths (Eds.), *Serengeti: dynamics of an ecosystem* (pp. 236–286). Chicago: University of Chicago Press

Houston, D. C. (1987). Management techniques for vultures—feeding and releases. In D. J. Hill (Ed.), *Breeding and management of birds of prey* (pp. 15–29). Bristol: University of Bristol.

Jaksic, F. M. (1981). Abuse and misuse of the term "guild" in ecological studies. *Oikos, 37,* 397–400.

Jedrzejewski, W., Zalewski, A., & Jedrzejewska, B. (1993). Foraging by pine marten (*Martes martes*) in relation tofood resources in Białowieza National Park, Poland. *Acta Theriologica, 38,* 405–426.

König, C. (1974). Zum verhalten spanischer Geier an Kadavern. *Journal für Ornithologie, 115,* 289–320.

König, C. (1983). Interspecific and intraspecific competition for food among old world vultures. In S. R. Wilbur & J. A. Jackson (Eds.), *Vulture biology and management* (pp. 153–171). Berkeley: University of California Press.

Kruuk, H. (1967). Competition for food between vultures in east Africa. *Ardea, 55,* 171–193.

López-López,P. García-Ripollés, C., & Urios, V (2014). Food predictability determines space use of endangered vultures: implications for management of supplementary feeding. *Ecological Applications 24,* 938–949.

Maes, J., Teller, A., Erhard, M., Liquete, C., Braat, L., et al. (2013). *Mapping and Assessment of Ecosystems and their Services. An analytical framework for ecosystem assessments under action 5 of the EU biodiversity strategy to 2020.* Publications office of the European Union, Luxembourg.

Margalida, A., & Heredia, R. (2005). *Biología de la conservación del quebrantahuesos (Gypaetus barbatus) en España. Organismo Autónomo Parques Nacionales.* Madrid: Organismo Autónomo De Parques Nacionales.

Margalida, A., Donázar, J. A., Carrete, M., & Sánchez-Zapata, J. A. (2010). Sanitary versus environmental policies: Fitting together two pieces of the puzzle of European vulture conservation. *Journal of Applied Ecology, 47,* 931–935.

Margalida, A., Colomer, M. A., & Sanuy, D. (2011). Can Wild Ungulate Carcasses Provide Enough Biomass to Maintain Avian Scavenger Populations? An Empirical Assessment Using a Bio-Inspired Computational Model. *PLoS ONE, 6*(5), e20248.

Margalida, A., Carrete, M., Hegglin, D., Serrano, D., Arenas, R., & Donázar, J. A. (2013). Uneven large-scale movement patterns in wild and reintroduced pre-adult bearded vultures: Conservation implications. PLoS ONE, *11*(8), e65857.

Margalida, A. Campión, D., & Donázar, J.A. (2014). Vultures vs livestock: conservation relationships in an emerging conflict between humans and wildlife. *Oryx, 48,* 172–176.

Markandya, A., Taylor, T., Longo, A., Murty, M. N., Murty, S., Dhavala, K. (2008). Counting the cost of vulture decline-An appraisal of the human health and other benefits of vultures in India. *Ecological Economics, 67,* 194–204.

Mateo, R., Rodríguez-de la Cruz M., Vidal, D., Reglero, M., & Camarero, P. (2007). Transfer of lead from shot pellets to game meat during cooking. *Science of the Total Environment, 372,* 480–485.

McCarthy, M. A., & Possingham, H. P. (2007). Active adaptive management for conservation. *Conservation Biology, 21,* 956–963.

Melis, C., Selva, N., Teurlings, I., Skarpe, C., Linnell, J. D. C., & Andersen, R. (2007). Soil and vegetation nutrient response to bison carcasses in Białowieza Primeval Forest, Poland. *Ecological Research, 22,* 807–813.

Merckx, T., & Pereira, H. M. (in press). Reshaping agri-environmental subsidies: From marginal farming to large-scale rewilding. *Basic and Applied Ecology.* doi: 10.1016/j.baae.2014.12.003

Milner, J. M., Bonenfant, C., Mysterud, A., Gaillard, J-M, Csányi, S., & Stenseth, N. C. (2006). Temporal and spatial development of red deer harvesting in Europe: biological and cultural factors. *Journal of Applied Ecology, 43,* 721–734.

Moreno-Opo, R. (2007). El buitre negro. In: R. Moreno-Opo, F. Guil (Coords.). *Manual de gestión del hábitat y las poblaciones de buitre negro en España* (pp. 25–45, 404 pp.). Dirección General para la Biodiversidad. Ministerio de Medio Ambiente. Madrid.

Moreno-Opo, R., Arredondo, A., & Guil, F. (2010). Foraging range and diet of Cinereous vulture *Aegypius monachus* using livestock resources in Central Spain. *Ardeola, 57,* 111–119.

Mundy, P., Butchart, D., Ledger, D., & Piper, S. (1992) *The vultures of Africa.* San Diego: Academy.

Olea, P. P., & Mateo-Tomás, P. (2009). The role of traditional farming practices in ecosystem conservation: The case of transhumance and vultures. *Biological Conservation, 142,* 1844–1853.

Oro, D., Margalida, A., Carrete, M., Heredia, R., & Donázar, J. A. (2008). Testing the goodness of supplementary feeding to enhance population viability of an endangered vulture. *PLoS ONE, 3*(12), e4084.

Ostfeld, R. S., & Keesing, F. (2000). Pulsed resources and community dynamics of consumers in terrestrial ecosystems. *Trends in Ecology and Evolution, 15,* 232–237.

Pain, D. J., & Pienkowski, M. W. (Eds.). (1997). *Farming and birds in Europe: The Common Agricultural Policy and its implications for bird conservation.* London: Academic.

Payne, J. A. (1965). A summer carrion study of the baby pig (*Sus scrofa,* Linnaeus). *Ecology, 46,* 592–602.

Piper, S. E. (2006). Supplementary feeding programs: How necessary are they for the maintenance of numerous and healthy vultures populations? In D. C. Houston & S. E. Piper (Eds.), *Proceedings of the international conference on conservation and management of vulture populations* (pp. 41–50). Natural History Museum of Crete WWF Greece.

Possingham, H. P., Andelman, S. J., Noon, B. R., Trombulak, S., & Pulliam, H. R. (2001). Making smart conservation decisions. In G. Orians & M. Soulé (Eds.), *Research priorities for conservation biology* (pp. 225–244). Washington, DC: Island Press.

Prakash, V., Bishwakarma, M. C., Chaudhary, A., Cuthbert, R., Dave, R., Kulkarni, M., Kumar, S., Paudel, K., Ranade, S., Shringarpure, R., & Green, R. (2012). The population decline of *Gyps* vultures in India and Nepal has slowed since the veterinary use of diclofenac was banned. *PLoS ONE, 7,* e49118.

Root, R. B. (1967). The niche exploitation pattern of the blue-grey gnatcatcher. *Ecological Monographs, 37,* 317–350.

Rose, M. D., & Polis, G. A. (1998). The distribution and abundance of coyotes: The effect of allochthonous food subsidies from the sea. *Ecology, 79,* 998–1007.

Sánchez-Zapata, J. A., Eguía, S., Blázquez, M., Moleón, M., & Botella, F. (2010). Unexpected role of ungulate carcasses in the diet of Golden Eagles *Aquila chrysaetos* in Mediterranean mountains. *Bird Study, 57,* 352–360.

Schluter, D., & Ricklefs, R. E. (1993). Species diversity. An introduction to the problem. In R. E. Ricklefs & D. Schluter (Eds.), *Species diversity in ecological communities* (pp. 1–10). Chicago: The University of Chicago Press.

Selva, N. (2004). The Role of Scavenging in the Predator Community of Bialowieza Primeval Forest. Tesis doctoral, Universidad de Sevilla, Sevilla, Spain.

Selva, N., Cortés-Avizanda, A. (2009). The effects of carcass and carrion dump site on communities and ecosystems. In J. A. Donázar, A. Margalida, & D. Campión (Eds.), *Vultures Feeding Stations and Sanitary Legislation: a Conflict and its Consequences from the Perspective of Conservation Biology* (pp. 452–473) Munibe 29 (suppl.), Sociedad de Ciencias Aranzadi, San Sebastián, Spain.

Selva, N., & Fortuna, M. A. (2007). The nested structure of a scavenger community. *Proceedings of the Royal Society of London B, 274,* 1101–1108.

Selva, N., Jedrzejewska, B., Jedrzejewski, W., & Wajrak, A. (2003). Scavenging on European bison carcasses in Białowieza Primeval Forest (eastern Poland). *Ecoscience, 10,* 303–311.

Selva, N., Jedrzejewska, B., Jedrzejewski, W., & Wajrak, A. (2005). Factors affecting carcass use by a guild of scavengers in European temperate woodland. *Canadian Journal of Zoology, 83,* 1590–1601.

Sikes, D. S. (1994). Influence of ungulate carcasses on coleopteran communities in Yellowstone National Park, USA. MSc Thesis, Montana State University, Montana, USA.

Slotta-Bachmayr, L., Bögel, R., & Camiña Cardenal, A. (Eds.). (2004). *The Eurasian Griffon Vulture (Gyps fulvus); in Europe and the Mediterranean: Status report and action plan.* East European/Mediterranean Vulture Working Group.

Towne, E. G. (2000). Prairie vegetation and soil nutrient responses to ungulate carcasses. *Oecologia, 122,* 232–239.

Tylianakis, J. M., Laliberté, E., Nielsen, A., & Bascompte, J. (2010). Conservation of species interaction networks. *Biological Conservation, 143,* 2270–2279.

Viñuela, J., Martí, R., & Ruiz, A. (1999). *El milano real en España. Monografía n_6.* SEO/BidLife. Madrid.

Wilbur, S. R., & Jackson, J. A. (1983). *Vulture biology and management.* Berkeley: University of California Press.

Wilmers, C. C., & Getz, W. M. (2005). Gray wolves as climate change buffers in Yellowstone. *PLoS Biology, 3*(4), e92.

Wilmers, C. C., & Post, E. (2006). Predicting the influence of wolf-provided carrion on community dynamics under climate change scenarios. *Global Change Biology, 12,* 403–409.

Wilmers, C. C., Stahler, D. R., Crabtree, R. L., Smith, D., & Getz, W. M. (2003). Resource dispersion and consumer dominance: scavenging at wolf and hunter-killed carcasses in Greater Yellowstone, USA. *Ecology Letters, 6,* 996–1003.

Yang, L. H., Bastow, J. L., Spence, K. O., & Wright, A. N. (2008). What can we learn from resource pulses? *Ecology, 89,* 621–634.

Chapter 6
Rewilding: Pitfalls and Opportunities for Moths and Butterflies

Thomas Merckx

Abstract Small organisms provide the bulk of biodiversity. Here, we look at rewilding from their perspective. As an umbrella group for other terrestrial invertebrates, we focus on the diverse group of Lepidoptera. More specifically, we set out to explore their response to farmland abandonment. So far, studies have warned against farmland abandonment, which is for instance listed as one of the key threats to European butterfly diversity. Here, partly based on a case study within the Peneda mountain range, we argue (i) that the majority of Lepidoptera is to a greater or lesser extent forest-dependent, (ii) that effects on species composition should be considered at regional rather than smaller scales, and (iii) that habitat resource heterogeneity at multiple spatial scales is key. As such, we believe that rewilding does offer opportunities to Lepidoptera. However, we recommend rewilding not to be equalled to a hands-off approach, but rather to a goal-driven conservation management approach. It should monitor, and where necessary intervene to provide habitat heterogeneity at multiple spatial scales, in order to cater for the whole gradient of sedentary to mobile species. Given that sufficient levels of habitat heterogeneity are maintained, Lepidoptera are one of probably many taxa that are likely to benefit from rewilding processes on European marginal farmland. The resulting improved species composition will help achieve European species conservation targets. It may also lead to more viable populations of moths, butterflies and other invertebrates, which will foster more resilient food-webs and increased ecosystem functioning.

Keywords Farmland abandonment · Habitat resource heterogeneity · Spatial scale · Controlled Rewilding · Lepidoptera

T. Merckx (✉)
Behavioural Ecology and Conservation Group, Biodiversity Research Centre, Earth and Life Institute, Université catholique de Louvain (UCL), Croix du Sud 4-5, bte. L7.07.04, 1348 Louvain-la-Neuve, Belgium
e-mail: th.merckx@gmail.com

Centro de Biologia Ambiental, Faculdade de Ciências da, Universidade de Lisboa, Campo Grande, 1749-016 Lisboa, Portugal

H. M. Pereira, L. M. Navarro (eds.), *Rewilding European Landscapes,*
DOI 10.1007/978-3-319-12039-3_6, © The Author(s) 2015

6.1   Rewilding Small-Sized Biodiversity Too

So far, the debate on rewilding opportunities for biodiversity has been mainly centred upon popular and hence large-sized taxa, such as large mammals and birds (see
Chaps. 1, 4, 5, and 8). As home range extent is typically mirrored by organismal
size, the relatively high mobility and large spatial footprint of large-sized taxa are
mainly situated at the extreme end of a whole gradient. The bulk of biodiversity is
smaller, less mobile, and hence operates at smaller spatial scales. Because of the
considerable dimension of the amount of European agricultural land that is already
being abandoned, and that is set to be abandoned over the next couple of decades
(see Chap. 1), it is obvious why the rewilding concept provides a most welcome
opportunity for wide-ranging and cursorial species, like wolves for example, whose
ecology has simply not been compatible with the typically small nature reserves
and the intensified countryside of Western Europe. However, for rewilding to be
adopted as a credible land-use option and conservation strategy, it will need to provide more than only a handful of iconic large animals. The successful uptake of the
rewilding approach may depend upon three main points: (i) rewilding will need
to make (socio-)economic sense (see Chap. 10), (ii) ample supply of ecosystem
services will need to be guaranteed (see Chap. 3), and last but not least, (iii) rewilding will need to make sense from a biodiversity conservation viewpoint too. As
such, it needs to cater for all kinds of biodiversity, i.e. for rare, range-restricted and
ubiquitous species, for generalists and specialists, for currently threatened and least-
concern species, for species operating at all kinds of spatial scales. With regard to
this latter point, we here look at rewilding from the perspective of smaller-sized
taxa. Although these taxa provide the bulk of biodiversity and ecosystem functioning, they are severely under-represented in conservation research (Clark and May
2002; Cardoso et al. 2011; Pereira et al. 2012). As an (incomplete) umbrella group
for other terrestrial invertebrates (Thomas 2005), we focus on the ecologically diverse group of Lepidoptera.

6.2   European Lepidoptera: Numbers and Trends

Lepidoptera are scale-winged and almost exclusively phytophagous insects, representing a mega-diverse radiation, probably correlated with the great diversification
of flowering plants since the Cretaceous (Menken et al. 2010). This major insect
order has been divided traditionally into the day-flying butterflies and largely nocturnal moths. In Europe, the order currently contains close to 9900 recorded species
(www.lepidoptera.pl), of which 482 species (i.e. ca. 5 %) are butterflies. About a
third of these butterfly species have currently declining populations on a European
scale, and 9 % are threatened (van Swaay et al. 2010). In some European regions
these figures are far higher; in Flanders, for example, 19 out of 67 resident butterfly
species (28 %) went extinct since the start of the twentieth century, whilst 25 species

(37%) are currently threatened (Maes et al. 2013). Such high proportions can be explained by cumulative effects of environmental pressures due to a long history of economic development (Dullinger et al. 2013). European-wide declines are especially worrisome since Lepidoptera provide many vital and economically important services within terrestrial ecosystems, such as nutrient cycling, prey resources and pollination. The European Red List for butterflies identifies the main drivers for these declines as habitat (connectivity) loss and degradation due to agricultural intensification and the invasion of shrubs and trees resulting from farmland abandonment (van Swaay et al. 2010).

Moth trends have not yet been evaluated at the European scale, although national trends are available for a handful of countries. For example, the very recent assessments in Britain and The Netherlands show a picture similar to butterflies. Across Britain, overall abundance of macro-moths declined by 28% over a recent 40-year period, with total numbers having decreased by 40% in the more populated southern half of Britain. Two-thirds of common and widespread macro-moth species for which national population trends were calculated, decreased in abundance, with 61 species having declined by 75% or more over 40 years (Fox et al. 2013). The preliminary Red List for macro-moths of The Netherlands (841 species) shows that 70 species (8%) went extinct since the nineteenth Century and that 300 species (35%) are currently threatened (Ellis et al. 2013). The decreasing extent of habitat and the degradation of its quality, more specifically via agricultural intensification, changing woodland management, urbanisation, climate change and light pollution, are likely causes of the observed changes in moth biodiversity (Fox et al. 2013).

6.3 Lepidoptera: Diurnal and Nocturnal Life-Styles

Thus, whilst farmland abandonment is generally reported to be relevant with regard to butterfly declines, moth declines are rarely linked to farmland abandonment. This observation makes sense if we consider their contrasting life-histories. Butterflies are day-flying ectotherms that need direct sunlight in order to raise their flight muscle temperature to optimal levels, most often well above ambient temperatures. By contrast, most moths are nocturnal and endothermic, raising their body temperatures above ambient levels by generating internal heat via muscle activity. As such, most butterflies and (ectothermic) day-flying moths occur typically in open, sun-lit habitats, whereas a majority of moths are to a greater or smaller degree linked to wooded habitats. For example, even in a largely deforested region (woodland cover <5%) of Flanders, 58% of in total 499 macro-moth species observed in 1980–2012 use shrubs or trees as foodplants (Sierens and Van de Kerckhove 2014). However, the situation is far from black and white as some butterflies are woodland specialists and many nocturnal moths require more open habitat conditions. Still, based on their contrasting thermoregulatory requirements it is clear why—in general—butterflies are perceived to be more susceptible than moths to farmland abandonment, which is typically accompanied by scrub and forest encroachment, shading out formerly sun-lit biotopes (van Swaay et al. 2010).

Although day-flying Lepidoptera are numerically the exception to the nocturnal norm, sound conservation strategies need to be inclusive of both. For example, in a context of temperate landscapes under intense human land-use, Merckx et al. (2012a) recommend zoned woodland management for the effective conservation of both life-history strategies. Their research showed that the late-successional deciduous woodland biotope is characterised by high numbers of both individuals and species of moths, being especially important for some scarce and specialist species of conservation concern (see also Baur et al. 2006), while coppicing and ride widening, which open up dense forest structures, are valuable woodland conservation tools for Lepidoptera species with an affinity for more open biotopes (see also Fartmann et al. 2013). The mechanism behind the pattern of increased lepidopteran species richness at the woodland-scale, due to such zoned management, involves an increased structural and micro-climatic diversity, and, more generally, increased habitat resource diversity (Merckx et al. 2012a). The importance of habitat heterogeneity at such larger scales is further highlighted by the results obtained from a recent study on butterfly richness of semi-natural meadows in Estonia; using a sample of 22 meadows with a total of 56 butterfly species, the research showed a positive correlation between butterfly diversity on local meadows and forest cover in the landscape directly surrounding (i.e. 250 m radius) these meadows, whilst meadow cover in the surrounding landscape at various spatial scales actually impacted the butterfly diversity of these meadows in a negative way (Ave Liivamägi pers. comm.). For the Mediterranean region too, Verdasca et al. (2012) showed that whilst regular management (i.e. removing understory vegetation) in oak stands has a positive effect on butterfly assemblages, undisturbed stands are nevertheless needed by some butterfly species. A study by Baur et al. (2006) on the abandonment of subalpine semi-natural grasslands in Transylvania, Romania, found that whilst vascular plants reached highest species richness in yearly-mown hay meadows, diurnal Lepidoptera were actually most species-rich in meadows abandoned for three years or more (see also Dover et al. 2011 and references therein), and nocturnal Lepidoptera and Gastropoda were most species-rich in young (20–50 years) and mature (50–100 years) forests, respectively. Because the complementarity of species composition increased with successional age in all four taxonomic groups, and because the proportion of red-listed nocturnal Lepidoptera increased with successional age too, their results indicate the high conservation value of all stages of grassland succession, and especially so the later seral stages up to mature woodland. Hence, all these studies highlight that although semi-natural open biotopes may locally reach high diversity levels (for some taxa), their partial abandonment at the landscape-scale has a role to play for biodiversity in general.

6.4   Conservation Objectives: Semi-Natural Biotopes Versus Rewilding

Conservation objectives are the subject of much debate for regions with a long history of human alteration, like in most of Europe (Merckx et al. 2013). Climax forests have been replaced, often millennia ago, by so-called semi-natural biotopes, which are essentially different versions of early- to mid-successional natural seral stages, arrested from developing towards mature woodland. Nowadays, only scattered fragments of ancient woodland remain, and these have suffered continuous but varying disturbance regimes by humans. For example, up to a century ago most European woodland was maintained as coppice or coppice-with-standards. Today much woodland plantation has a uniformly closed canopy, which is shadier than is found in ancient forests with little history of human disturbance. A consequence of these various ways by which the development towards climax forest has been arrested, is that the rarest types of arboreal habitats in west-European countries today are those associated with rotting wood on ancient trees. These saproxylic habitats are associated with high extinction rates (Hambler et al. 2011), and support numerous invertebrate specialists, especially beetles and flies, and some moths (Thomas et al. 1994). Although many species undoubtedly suffered great losses (and extinctions) over the centuries during the transition from natural forest to semi-natural habitats, others (e.g. species associated with early-successional stages) did benefit or managed to adapt successfully to these semi-natural biotopes (e.g. heaths, meadows, coppiced woodland) (Young 1997; Monbiot 2013).

Since the 1950s, agricultural intensification, forestry intensification, urbanization, and farmland abandonment have all severely decreased such semi-natural biotopes in quantity, quality, and connectivity, and with them their specialist fauna and Lepidoptera. As a result, most European 'conservationists' traditionally seek to sustain or restore semi-natural biotopes, and do so by maintaining very specific disturbance regimes, often simply by copying traditional agricultural practices, since most large wild herbivores were excluded centuries ago (New 2009). Nevertheless, popular management operations to influence vegetation structure, such as burning and grazing/mowing, need careful planning (mainly to ensure refugia) as they may destroy much of the existing invertebrate fauna if applied too intensively, too infrequently, on too large a scale or at unsuitable times of year (New 2009). Hence, although wrongly applied conservation management may have unintended negative consequences, management is seen as a good thing overall, whereas rewilding is often perceived as a threat (Merckx et al. 2013).

Abandonment of human disturbance does indeed pose a threat to many specialist species that have become adapted to certain semi-natural biotopes, especially when they have nowhere else to go because natural succession dynamics are currently too disturbed and suitable natural patches are too small and/or isolated. On the other hand, the recovery of native forest ecosystems due to farmland abandonment is likely to benefit a majority of Lepidoptera, since most moths (which make up 95% of Lepidoptera) are reliant on woody foodplants or wooded biotopes. Forest recovery will obviously favour endangered specialist faunal groups too (e.g. closed-woodland Lepidoptera, Gastropoda, and saproxylic groups).

6.5 Controlled Rewilding: Reconciling the Objectives

So, how do we reconcile things? Earlier, we reported on a woodland conservation management policy inclusive of both open- and closed-habitat species at the woodland patch scale (Merckx et al. 2012a). Here, we propose another win-win situation, but now within a rewilding context: a conservation management strategy that could reconcile the needs of semi-natural habitat specialists at the regional scale.

The outcome of a management strategy is likely to differ dependent on whether it is applied at the local, landscape or regional scale, and this because species diversity under conditions of low land-use intensity is strongly dependent on spatial scale (see Chap. 1, Fig. 1.2). We here opt for the regional scale, because we believe that within a European context it is more important, and more feasible too, to safeguard a certain species somewhere within a certain region (i.e. regional scale), rather than to safeguard the precise locations of the local populations of that species (i.e. local scale). As such, we consider the relevant spatial scale at which to consider the effects on species composition of the proposed strategy to be the regional (i.e. ca. 100×100 km) scale.

Our proposed strategy is not intended to be applied to European regions with fertile agricultural land, but rather to regions characterised by agricultural marginal land. Merckx and Pereira (in press) already warned against an overly agro-centric view on conservation for marginal land, which instead does provide excellent rewilding opportunities. Under our strategy, which we call 'Controlled Rewilding'—combining forest recovery with monitoring and management of semi-natural biotopes—many such regions could evolve towards mosaics including mature climax vegetation, semi-natural biotopes, and natural successional stages, such as river areas, wood gaps and high-altitude areas. Nevertheless, open habitats may be rarer than in pre-historic landscapes, owing to the absence of most former natural herbivores (Merckx et al. 2013), but sun-lit biotopes and grazing can be achieved via other means (see Chap. 8). Eventually, the recovered forests within the resulting mosaics will become more and more dynamic and heterogeneous as a result of natural disturbance regimes operating on a wide range of spatial scales, characteristic of natural forests, and which further enhance biodiversity (Lindenmayer et al. 2006; Lehnert et al. 2013).

We believe there is room within Europe for Controlled Rewilding. It entails both passive abandonment and active (temporary) management interventions to ecologically restore semi-natural biotopes within a rewilding context. As a prerequisite, this strategy needs to include the monitoring of habitat heterogeneity levels at multiple spatial scales. Its aim is to pinpoint where conservation management interventions are required, so as to provide sufficient levels of habitat heterogeneity for specialists of both open and closed biotopes, and at multiple spatial scales in order to cater for

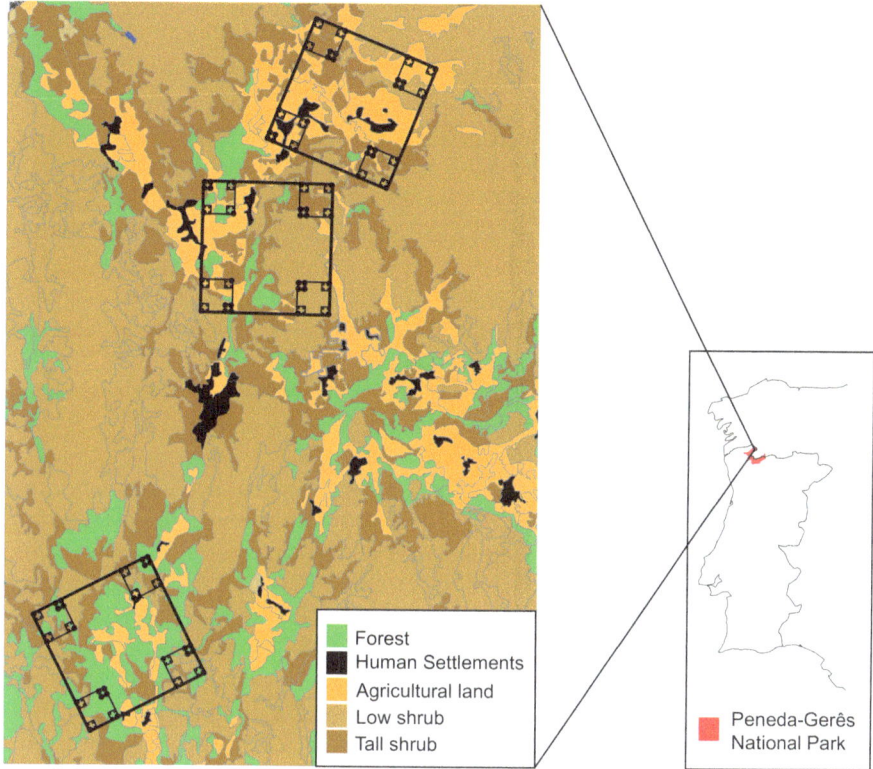

Fig. 6.1 Location of the study area within Portugal. Semi-nested sampling design: Each black dot represents the location of a fixed light-trap sampling site. Sampling occurred at four spatial scales (i.e. 20×20 m; 80×80 m; 320×320 m; 1280×1280 m) within each of three 'landscapes' that differed in terms of their dominant land-use cover: (i) 'Forest-dominated': the southernmost 'landscape' was mainly characterised by native, semi-natural forest, (ii) 'Shrub-dominated': the middle 'landscape' was mainly characterised by scrubland, and (iii) 'Meadow-dominated': the northernmost 'landscape' was mainly characterised by extensively managed meadows

small-sized (less mobile) and large-sized taxa (see above). In our opinion, conservation focused on semi-natural biotopes and rewilding should be complementary, via a Controlled Rewilding strategy, and now is the time to designate regions within Europe where both approaches could be combined (see Chaps. 2 and 11).

6.6   What About Fertile Agricultural Regions?

Does Controlled Rewilding for marginal land mean that we should forget about the ecosystems within fertile agricultural regions? Although it makes sense within a Europe-wide land-sparing framework to intensively farm such fertile regions (Merckx and Pereira in press), we remark that a dominant land-use of intensive farming does not necessarily have to imply current destructive practices, but merely the provision of high yields. Moreover, the conservation value of remaining semi-natural patches can be increased by an ecological upgrade of the intervening 'matrix', which basically consists of farmland, but also of brownfield sites and urbanised areas (Dennis 2010). Tangible environmental benefits can be obtained on farmland via agri-environment schemes (AES) (Donald and Evans 2006; Scheper et al. 2013), where the aim should be to reconcile intensive agricultural practices with wider societal benefits, including biodiversity. Here, the basic questions are which landscape elements to restore, how, and at what spatial scale, in order to make farmland less hostile to a broad range of declining 'wider countryside' and rare, localized species (Merckx et al. 2010a). Brownfield and even urban sites provide opportunities for restoration of successional biotopes otherwise not strongly represented locally, with restoration plans best tailored to focal species and/or to improving biotopes by assuring a sufficient quantity, quality and spatio-temporal diversity of habitat resources (New 2009; Dennis 2010).

AES can reverse negative biodiversity trends by increasing resource heterogeneity and improving dispersal success (Dennis 2010). However, they must be made more efficient and cost-effective (Scheper et al. 2013). One way to achieve this is by implementing specific measures for high-priority species targeted at landscapes where such species occur. However, we argue that this species-specific approach must be complemented by a multi-species approach in order to more fully address the steep declines in farmland biodiversity. General AES that are focused on the restoration and implementation of vital landscape elements are key to this multi-species approach. Even simple AES management prescriptions applied to relatively small areas can benefit Lepidoptera populations. For example, the restoration and management of arable field margins has been shown to benefit a range of insect groups (Haaland et al. 2011). Sympathetic management of hedgerows has positive effects on vulnerable insects, such as the brown hairstreak *Thecla betulae* butterfly (Merckx and Berwaerts 2010) and the lackey *Malacosoma neustria* (T. Merckx pers. data), a macro-moth of which the larvae feed gregariously on blackthorn *Prunus spinosa* and hawthorn *Crataegus* sp. In addition, we have recently discovered that the protection of existing hedgerow trees, and the provision of new ones, is likely to be a highly beneficial conservation tool for populations of moths, and probably many other flying insects too, as hedgerow trees provide a sheltered microclimate and other key habitat resources (Merckx et al. 2012b). The implementation of hedgerow tree and field margin AES options is likely to provide even better results in areas where farmers are targeted to join AES across the landscape. This approach results in a landscape-scale joining-up of habitat resources. Such a higher connectivity between resources at the landscape-scale does benefit fairly mobile species, which use the farmland biotope at a landscape-scale (see Chap. 7). A fair amount of macro-moth species falls into this category (Merckx and Macdonald in press).

6.7   A Case Study: Farmland Abandonment in Peneda and Its Effects on Macro-moths

Coming back to the European marginal land context, where farmland abandonment is currently perceived as a major threat to the diversity of specialist butterflies from open biotopes and to other open-biotope invertebrates (Marini et al. 2009a; van Swaay et al. 2010), we recently carried out a two-year project (2011–2012) in Portugal to study the response of moths to farmland abandonment. There are a total of 2583 Lepidoptera species currently recognised from Portugal (Corley et al. 2013). We light-trapped macro-moths at 84 fixed sampling sites, each of which was repeatedly sampled six times. Sampling sites were part of a semi-nested sampling design (Proença and Pereira 2013) in three study landscapes that represented a farmland abandonment gradient within the Peneda mountain range (NW-Portugal; Fig. 6.1; elevation 750–1155 m). Here, an on-going rural exodus since the 1950s has led to farmland abandonment and regeneration of native woodland cover, although significant amounts of land, especially close to villages, are still being farmed (Rodrigues 2010). This situation is common to many other rural areas in Europe, and the Peneda area is considered to be representative of other traditional agricultural landscapes in mountainous areas of Southern Europe (Queiroz et al. submitted). We were able to analyse the species composition response to farmland abandonment and forest regeneration of both open-biotope and closed-biotope species across spatial scales, from local up to landscape scales. Nice add-ons to our research were the discovery of a micro-moth species new to science, namely *Isotrias penedana*, currently only known from the Peneda mountain range (Trematerra 2013; Fig. 6.2a), as well as the discovery of 12 macro-moth species new for Portugal (Corley et al. 2012, 2013; Fig. 6.2b).

a   b

Fig. 6.2 a Notable moth discoveries *Isotrias penedana*, a tortricid (micro-moth) species new to science, discovered during 2012 within the Peneda mountain range (Trematerra 2013) (photo credit: Martin Corley). **b** *Watsonarctia deserta*, an arctiid (macro-moth) species added to the Portuguese list in 2012. In Portugal currently only known from four locations, all within the Peneda mountain range (Corley et al. 2013). (Photo credit: Eduardo Marabuto)

In a first analysis of the ca. 23.000 individuals and ca. 380 species rich dataset, we lumped the data from the six sampling rounds for each of the 84 local sites (20×20 m, as light traps have relatively small attraction radii: Merckx and Slade 2014). This analysis shows that overall macro-moth species richness is locally richest at woodland sites (mean $\pm$ SE: 77.1 ± 4.2), intermediate at meadows (63.2 ± 3.8) and tall shrub (61.7 ± 5.9), and poorest at low shrub sites (51.0 ± 4.3). Not only species richness, but functional diversity too was significantly higher within woodland than in meadows or shrubland for macro-moths (Queiroz et al. submitted). Although this overall pattern among the four main biotopes is basically present in the meadow-dominated, shrub-dominated and forest-dominated landscapes, species richness is locally, at all four biotope types, consistently lower within the shrub-dominated landscape than in meadow-dominated and forest-dominated landscapes (Fig. 6.3a). Possible confounding factors, such as altitude, slope and soil fertility, were not analysed. Although they may have played a role in the patterns observed, their role is likely to have been small given the large amount of sites sampled within the same region, and given the limited altitudinal differences between sites.

These observations suggest that local abandonment of relatively species-rich, semi-natural meadows may reduce (-20%) local richness levels of nocturnal macro-moths, when meadows turn into low shrub biotopes, but that richness levels should locally increase again when these abandoned fields gradually turn into native forest, eventually reaching diversity levels well above those associated with meadows ($+22\%$). The above results may also mean that when the resulting shrubby vegetation from farmland abandonment becomes the dominant land cover within a landscape (for instance due to simultaneous abandonment of large areas within the same landscape and/or due to the arrested development of the resulting low shrub by overgrazing/shepherding and fire), that this landscape effect lowers local species richness levels. This would happen within remaining meadows and woodlands due to fragmentation effects, and also within the scrubland itself due to lower species inputs from different habitats (i.e. reduced spillover effects because of lower heterogeneity). Eventually, such a process may lead to an impoverished species composition of mainly shrub specialists (e.g. the horse chestnut moth *Pachycnemia hippocastanaria*) and ubiquitous species, and this at both the local and landscape-scale.

Our data also show that this negative landscape-effect of a dominance of shrub vegetation within the surrounding landscape does not only reduce local species richness, affecting all biotopes, but translates to reduced species richness at larger scales too (Fig. 6.3b). By contrast, overall macro-moth richness was highest in the forest-dominated landscape and intermediate in the meadow-dominated landscape, at all four spatial scales tested (Fig. 6.3b). Moreover, the difference between the forest-dominated and meadow-dominated landscape in absolute species richness steadily increases with spatial scale, from an 11 species difference at the local scale, over a 24 and 32 species difference at intermediate spatial scales, to an excess of 41 species for the forest-dominated landscape at the largest scale tested (i.e. 1280×1280 m) (Fig. 6.3b). We interpret these results as follows: extensively farmed landscapes do indeed provide high levels of moth diversity, both locally and at the landscape-scale. Still, more has to be gained from landscapes with a high amount of forest

Fig. 6.3 Macro-moth species richness ($\pm$ SE)—Local sites (20×20 m; $N=84$) were each sampled six times over 2 years (2011–2012). Data from the sampling rounds were aggregated for each site. **a** Overall richness at local sites for each of four biotopes (Meadow; Low Shrub; Tall Shrub; Woodland), separately for three landscape types (Meadow-dominated; Shrub-dominated; Forest-dominated) (see also Fig. 6.1). Note that in the forest-dominated landscape only one meadow site and two tall shrub sites were sampled, explaining the absent and very large error bars, respectively. **b** Overall richness at four spatial scales [20×20 m: local site ($N=84$); 80×80 m: lumping four sites ($N=12$); 320×320 m: lumping seven sites ($N=12$); 1280×1280 m: lumping 28 sites ($N=3$)], separately for three landscape types (Meadow-dominated; Shrub-dominated; Forest-dominated) (see also Fig. 6.1). The absence of error bars at the largest spatial scale is because these three landscape types were each represented by one landscape only. **c** Richness of closed-biotope (i.e. woodland) versus open-biotope species at five spatial scales (see Fig. 6.3b, but with an additional larger scale lumping all 84 sites). For contrast, a majority of species ($N=223$) that occur in mixed or intermediate biotopes are not retained

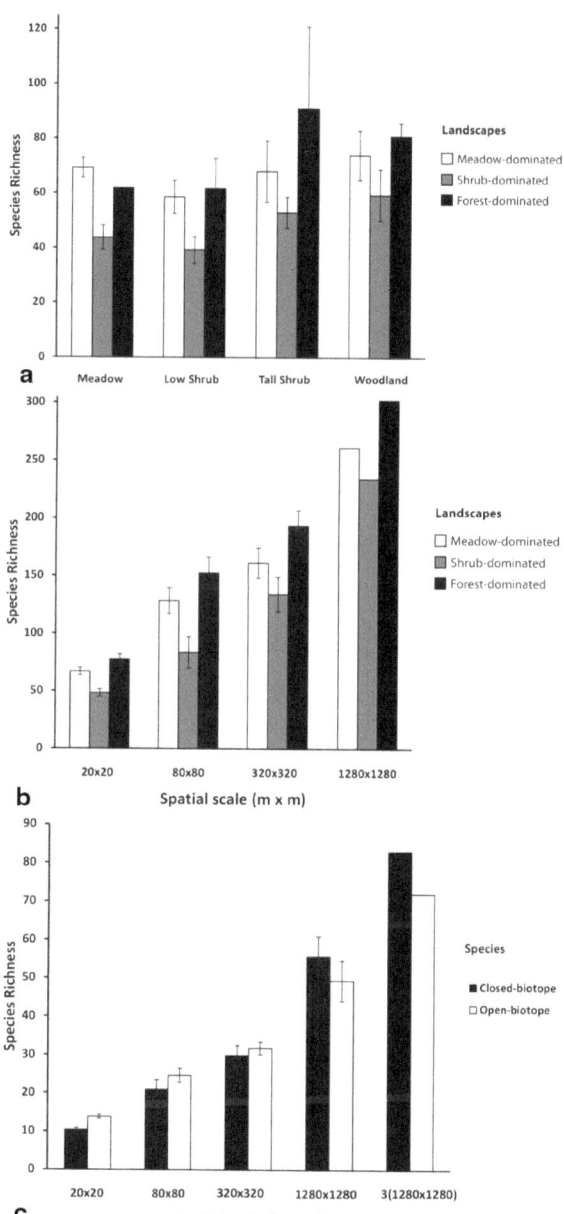

cover. Although the species richness difference is not that big at the local scale, the difference becomes larger and more notable with increasing scale, which points to larger beta-diversity levels for forest-dominated than for meadow-dominated landscapes/regions. Indeed, it is well known that semi-natural grasslands are able to reach high local species richness (i.e. alpha diversity), at least for certain taxa (e.g. butterflies: van Swaay 2002; flowering plants: Wilson et al. 2012). However, we show, for macro-moths, that those local diversity levels can be even higher within woodlands, and importantly, that beta-diversity levels may be consistently larger at larger spatial scales within landscapes mainly covered by forests compared to landscapes mainly covered by extensively managed agricultural land. So, although the effects on macro-moth diversity of woodlands and forest-dominated landscapes already compare positively with meadows and meadow-dominated landscapes at the local scale, forest-dominated landscapes outcompete meadow-dominated landscapes more strongly at larger spatial scales. We believe this is an important point since rewilding is to be applied at larger, regional scales, whereas biodiversity has traditionally been measured mainly at local scales alone.

A key result linked to this is the interaction we observed between spatial scale and species' biotope characteristics, whereby the species richness of woodland moths shows a steeper increase with spatial scale than the richness of open-biotope moths (Fig. 6.3c). Whilst we found on average more open-biotope species than typical woodland species locally, this difference disappears at the field-scale, and reverses at the largest scales. Thus, it appears that closed-biotope species have higher beta-diversity than open-biotope species, and that the former are indeed responsible for the higher diversity levels within forest-dominated landscapes.

Finally, these results also demonstrate that although forest-dominated landscapes provide advantages over meadow-dominated landscapes, one needs to take good care not to get stuck into a landscape largely dominated by shrubs alone (Fig. 6.3a, b). Rather, we advise to monitor, and if needed to cater for spatial and temporal habitat heterogeneity within landscapes undergoing farmland abandonment, so as to combat these negative effects.

6.8   Habitat Resource Heterogeneity at Multiple Spatial Scales is Key

Let's now zoom out from the specifics of the Peneda case study on macro-moths to a more general view on rewilding landscapes for multiple taxa. In order to do so, it is essential to consider dispersal, not only because it is a fundamental process that bridges across spatial scales (Chave 2013), but also because it allows us to understand the importance of habitat resource heterogeneity. Non-sessile organisms need to move in order to reach their essential habitat resources, needed for the completion of their life-cycle. These varied resources are often spread over spatial and temporal gradients, and are most often patchily distributed, even within continuous vegetation types (Dennis 2010). Since dispersal is costly in terms of energy expenditure and predation risk, organisms hence generally benefit from resource configurations that limit dispersal needs (Vanreusel and Van Dyck 2007). Landscapes characterised by high habitat resource heterogeneity are more likely to provide such resource configurations (Dennis 2010; Vickery and Arlettaz 2012), and hence to provide increased species diversity (Verhulst et al. 2004). However, since there is a vast variety in terms of dispersal capacities and resource use among, and even within taxa, such heterogeneity needs to be provided at multiple spatial scales in order to cater for all. After all, each species/individual is adapted to exploit resources within a spatial range, some more specifically so than others, and this over a whole gradient from extreme widely to extreme narrowly spaced resources. Evidently, the more biotope types within a landscape, the wider the array of resources and thus the more organisms supported. Also, for a given species, large enough quantities of its essential resources need to be present, limiting the minimum patch size below which it will be absent. In general, mobile organisms need larger areas of habitat than less-mobile ones (Pereira et al. 2004). For instance, mobile woodland moth species were not found in small (<5 ha) wood patches (Slade et al. 2013). As such, the high macro-moth diversity in the forest-dominated landscape of Peneda (see above) can be explained by a combination of (i) sufficiently large woodland patches to cater for the needs of mobile woodland specialists, and (ii) sufficient resource heterogeneity at multiple spatial scales. Indeed, the specific study landscape was not a forest 'monoculture' but consisted of a patchwork of meadows and scrubland of varying size embedded within a dominant forest matrix (Fig. 6.1). This patchwork allows a diverse composition of meadow, scrubland and woodland species.

Habitat heterogeneity is known to strongly influence the abundance and diversity of species within landscapes (Tjørve 2002). For instance, the change of the typically diverse habitat mosaic of extensive farmland towards the spatially and temporally—both at multiple scales—increasingly uniform intensive farmscapes, has been identified as the root cause of the decline in European farmland biodiversity, whether measured at a small or large scale (Benton et al. 2003). Likewise, we believe that rewilding exercises need to pay attention that sufficient heterogeneity remains during the whole rewilding process. For example, whilst bird species typical of extensive farmland disappear when landscapes become too open due to intensification, they may at the other extreme also disappear from areas where forest recovery removes all open areas (Vickery and Arlettaz 2012). Rewilding projects should hence monitor heterogeneity and intervene with conservation management when necessary (Controlled Rewilding: see above). Because the intense defaunation (especially of large mammals) since the Pleistocene has lead to a decreased environmental heterogeneity in remaining ecosystems (Corlett 2013), and since current rewilding exercises are not continent-wide but region-wide at best, it should be clear that hands-off restoration processes may not be sufficient to reach desirable biodiversity outcomes, and that Controlled Rewilding is more advisable.

6.9 Wrapping It Up and the Way Forward

Semi-natural meadows and other open biotopes are important contributors to agro-ecosystem biodiversity on marginal land (Knop et al. 2006). Whilst the negative effects of grassland intensification on biodiversity are relatively well understood

(Marini et al. 2008), the impact of grassland abandonment—common to European mountainous areas—is much less understood, and hence controversial. In our view, the alleged negative effects result from two biases: species composition studies are often performed at relatively small scales, and narrowly focus on open-biotope species alone. These biases ignore that ecosystem dynamics occur across spatial scales, and that ecosystems consist of other functional groups too.

Nevertheless, evidence exists that cessation of management, and the resulting transitional vegetation types, can be important—even at the field-scale—for several taxonomic groups (Lepidoptera: Balmer and Erhardt 2000; Baur et al. 2006; Öckinger et al. 2006; Skórka et al. 2007—Apoidea: Kruess and Tscharntke 2002—Orthoptera: Marini et al. 2009a). Similarly, a recent review paper convincingly shows that reduced grazing generally positively affects arthropod diversity (van Klink et al. 2014).

We believe that studies on biodiversity effects should include landscape and/or regional scales. Ecological processes are not limited to the field-scale, as many organisms move at larger spatial scales, and respond differently to the surrounding landscape according to their size, mobility and functional traits (Steffan-Dewenter et al. 2002; Tscharntke et al. 2005; Merckx et al. 2009, 2010b; Slade et al. 2013; Queiroz et al. submitted). At these larger scales we expect positive effects from land abandonment. Indeed, reduced effects of habitat fragmentation, and high landscape heterogeneity, may lead to better functioning, more diverse, and hence more resilient ecosystems (Loreau and de Mazancourt 2013). For example, Marini et al. (2008, 2009b) show that a high proportion of woody vegetation at a landscape-scale positively affects Orthopteran species richness, whereas a high proportion of grasslands did so negatively. Other examples are birds of prey which obtain resources from both farm- and woodland. Their densities tend to increase with forest cover; Booted eagle (*Aquila pennata*) densities, for instance, peak at ca. 80% forest cover (Sánchez-Zapata and Calvo 1999). Conservation of raptors, and many other mobile taxa, thus requires a regional approach towards the right landscape mosaics of forests and open biotopes (Vickery and Arlettaz 2012). A marked increase in native forest cover within Europe may locally also combat the increased frequency and intensity of fires due to climate change (Proença et al. 2010), which are considered threats to European butterfly diversity (van Swaay et al. 2010). Nevertheless, as the highest diversity of (threatened) butterflies is found in mountainous areas in Southern Europe, where numerous restricted-range species are encountered (van Swaay et al. 2010), and where a high degree of farmland abandonment is likely (see Chap. 1, Fig. 1.4), rewilding projects in such areas will need to make sure they retain all key habitat resources of such high-priority species at sufficient levels within (managed) open habitats (see Chap. 8).

Here, we reported on research within the Peneda mountain range in which we looked at the response of macro-moth species composition to farmland abandonment across multiple spatial scales. Based on our findings and on general ecological insights, we strongly recommend that a possible rewilding approach would not be equalled to a hands-off approach, but rather to a conservation management approach that monitors, and when necessary intervenes to provide habitat heterogeneity at multiple spatial scales, in order to cater for the whole gradient of sedentary to mobile species. Given that sufficient levels of habitat heterogeneity are maintained, Lepidoptera are one of probably many taxa that are likely to benefit from rewilding exercises on European marginal farmland. An important point to take on board for nocturnal Lepidoptera, and other nocturnal biodiversity, is the issue of light pollution, which may cause adverse effects on larval development, fitness and population dynamics (van Geffen et al. 2014), and may hence result in negative effects cascading through whole ecosystems (van Langevelde et al. 2011; Fox et al. 2013). Rewilding exercises should thus include measures to actively mitigate light pollution sources (see Chap. 2). The likely improved species composition as a result will not only help achieve European species conservation targets (i.e. halting and reversing declines in biodiversity), but stronger populations of moths, butterflies and other invertebrates will also result in increased resilience of food-webs and ecosystem functioning (Loreau and de Mazancourt 2013).

Since there is an urgent need to determine the effects of farmland abandonment and landscape context on communities over a wide range of spatial scales (Cozzi et al. 2008), we believe that our study on macro-moths in the Peneda range provides useful and timely scientific evidence on the benefits, requirements and limitations of the rewilding approach for marginal land, such as mountainous areas and High Nature Value farmland within Europe. We hope that the approach of Controlled Rewilding here proposed, may help design an effective conservation policy regarding the European farmland abandonment process. We here call for more research into how optimal configurations of rewilded land on abandoned farmland would look like. We also call to develop regional goals for biodiversity on land characterised by a high degree of farmland abandonment. Although it is clear that a universal approach to provide habitat heterogeneity at multiple spatial scales is the key to deliver and sustain the required biodiversity, we may need different types of heterogeneity in different places and regions.

Acknowledgements I am immensely grateful to Martin Corley, who provided lots of help and advice with the identification of the numerous samples. I also thank Henrique M. Pereira, Laetitia M. Navarro, the Theoretical Ecology and Biodiversity Change Group, Eduardo Marabuto, Anabela Moedas and Pedro Alarcão for their help and support throughout. The Fundação para a Ciência e a Technologia provided financial support (postdoc grant SFRH/BPD/74393/2010), and both the Peneda-Gerês National Park and the Instituto da Conservação da Natureza e da Biodiversidade allowed carrying out the fieldwork.

References

Balmer, O., & Erhardt, A. (2000). Consequences of succession on extensively grazed grasslands for Central European butterfly communities: Rethinking conservation practices. *Conservation Biology, 14,* 746–757.

Baur, B., Cremene, C., Groza, G., Rakosy, L., Schileyko, A. A., Baur, A., Stoll, P., & Erhardt, A. (2006). Effects of abandonment of subalpine hay meadows on plant and invertebrate diversity in Transylvania, Romania. *Biological Conservation, 132,* 261–273.

Benton, T. G., Vickery, J. A., & Wilson, J. D. (2003). Farmland biodiversity: Is habitat heterogeneity the key? *Trends in Ecology and Evolution, 18,* 182–188.

Cardoso, P., Erwin, T. L., Borges, P. A., & New, T. R. (2011). The seven impediments in invertebrate conservation and how to overcome them. *Biological Conservation, 144*, 2647–2655.

Chave, J. (2013). The problem of pattern and scale in ecology: What have we learned in 20 years? *Ecology Letters, 16*, 4–16.

Clark, J. A., & May, R. M. (2002). Taxonomic bias in conservation research. *Science, 297*, 191–192.

Corlett, R. T. (2013). The shifted baseline: Prehistoric defaunation in the tropics and its consequences for biodiversity conservation. *Biological Conservation, 163*, 13–21.

Corley, M. F. V., Merckx, T., Cardoso, J. P., Dale, M. J., Marabuto, E., Maravalhas, E., & Pires, P. (2012). New and interesting Portuguese Lepidoptera records from 2011 (Insecta: Lepidoptera). *SHILAP Revista de Lepidopterología, 40*, 489–511.

Corley, M. F. V., Merckx, T., Marabuto, E., Arnscheid, W., & Maravalhas, E. (2013) New and interesting Portuguese Lepidoptera records from 2012 (Insecta: Lepidoptera). *SHILAP Revista de Lepidopterología, 41*, 449–477.

Cozzi, G., Müller, C. B., & Krauss, J. (2008). How do local habitat management and landscape structure at different spatial scales affect fritillary butterfly distribution on fragmented wetlands? *Landscape Ecology, 23*, 269–283.

Dennis, R. L. H. (2010). *A resource-based habitat view for conservation: Butterflies in the British landscape.* Chichester: Wiley-Blackwell.

Donald, P. F., & Evans, A. D. (2006). Habitat connectivity and matrix restoration: The wider implications of agri-environment schemes. *Journal of Applied Ecology, 43*, 209–218.

Dover, J. W., Spencer, S., Collins, S., Hadjigeorgiou, I., & Rescia, A. (2011). Grassland butterflies and low intensity farming in Europe. *Journal of Insect Conservation, 15*, 129–137.

Dullinger, S., Essl, F., Rabitsch, W., Erb, K. H., Gingrich, S., Haberl, H., Hülber, K., Jarosik, V., Krausmann, F., Kühn, I., Pergl, J., Pysek, P., & Hulme, P. E. (2013). Europe's other debt crisis caused by the long legacy of future extinctions. *Proceedings of the National Academy of Sciences, 110*, 7342–7347.

Ellis, W., Groenendijk, D., Groenendijk, M., Huigens, T., Jansens, M., van der Meulen, J., van Nieukerken, E., & de Vos, R. (2013). *Nachtvlinders Belicht—Dynamisch, Belangrijk, Bedreigd.* De Vlinderstichting: The Netherlands.

Fartmann, T., Müller, C., & Poniatowski, D. (2013). Effects of coppicing on butterfly communities of woodlands. *Biological Conservation, 159*, 396–404.

Fox, R., Parsons, M. S., Chapman, J. W., Woiwod, I. P., Warren, M. S., & Brooks, D. R. (2013). The State of Britain's Larger Moths 2013. Butterfly Conservation and Rothamsted Research, Wareham, Dorset, UK.

Haaland, C., Naisbit, R. E., & Bersier, L. F. (2011). Sown wildflower strips for insect conservation: a review. *Insect Conservation and Diversity, 4*, 60–80.

Hambler, C., Henderson, P. A., & Speight, M. R. (2011). Extinction rates, extinction-prone habitats, and indicator groups in Britain and at larger scales. *Biological Conservation, 144*, 713–721.

Knop, E., Kleijn, D., Herzog, F., & Schmid, B. (2006). Effectiveness of the Swiss agri-environment scheme in promoting biodiversity. *Journal of Applied Ecology, 43*, 120–127.

Kruess, A., & Tscharntke, T. (2002). Grazing intensity and the diversity of grasshoppers, butterflies and trap-nesting bees and wasps. *Conservation Biology, 16*, 1570–1580.

Lehnert, L. W., Bässler, C., Brandl, R., Burton, P. J., & Müller, J. (2013). Conservation value of forests attacked by bark beetles: Highest number of indicator species is found in early successional stages. *Journal for Nature Conservation, 21*, 97–104.

Lindenmayer, D. B., Franklin, J. F., & Fischer, J. (2006). General management principles and a checklist of strategies to guide forest biodiversity conservation. *Biological Conservation, 131*, 433–445.

Loreau, M., & de Mazancourt, C. (2013). Biodiversity and ecosystem stability: A synthesis of underlying mechanisms. *Ecology Letters, 16*, 106–115.

Maes, D., Vanreusel, W., & Van Dyck, H. (2013). Dagvlinders in Vlaanderen—Nieuwe Kennis voor Betere Actie (Butterflies in Flanders—New Knowledge for Better Action). Lannoo, Tielt, Belgium.

Marini, L., Fontana, P., Scotton, M., & Klimek, S. (2008). Vascular plant and Orthoptera diversity in relation to grassland management and landscape composition in the European Alps. *Journal of Applied Ecology, 45*, 361–370.

Marini, L., Fontana, P., Battisti, A., & Gaston, K. J. (2009a). Response of orthopteran diversity to abandonment of semi-natural meadows. *Agriculture, Ecosystems and Environment, 132*, 232–236.

Marini, L., Fontana, P., Battisti, A., & Gaston, K. J. (2009b). Agricultural management, vegetation traits and landscape drive orthopteran and butterfly diversity in a grassland-forest mosaic: A multi-scale approach. *Insect Conservation and Diversity, 2*, 213–220.

Menken, S. B. J., Boomsma, J. J., & van Nieukerken, E. J. (2010). Large-scale evolutionary patterns of host plant associations in the Lepidoptera. *Evolution, 64*, 1098–1119.

Merckx, T., & Berwaerts, K. (2010). What type of hedgerows do Brown hairstreak (*Thecla betulae* L.) butterflies prefer? Implications for European agricultural landscape conservation. *Insect Conservation and Diversity, 3*, 194–204.

Merckx, T., & Macdonald, D. W. (in press). Landscape-scale conservation of farmland moths. In D. W. Macdonald & R. E. Feber (Eds.), *Wildlife conservation on farmland*. Oxford University Press.

Merckx, T. & Pereira, H. M. (in press). Reshaping agri-environmental subsidies: From marginal farming to large-scale rewilding. Basic and Applied Ecology. doi: 10.1016/j.baae.2014.12.003

Merckx, T., & Slade, E. M. (2014). Macro-moth families differ in their attraction to light: Implications for light trap monitoring programs. *Insect Conservation and Diversity, 7*, 453-461.

Merckx, T., Feber, R. E., Dulieu, R. L., Townsend, M. C., Parsons, M. S., Bourn, N. A. D., Riordan, P., & Macdonald, D. W. (2009). Effect of field margins on moths depends on species mobility: Field-based evidence for landscape-scale conservation. *Agriculture, Ecosystems and Environment, 129*, 302–309.

Merckx, T., Feber, R. E., Parsons, M. S., Bourn, N. A. D., Townsend, M. C., Riordan, P., & Macdonald, D. W. (2010a). Habitat preference and mobility of *Polia bombycina*: Are non-tailored agri-environment schemes any good for a rare and localised species? *Journal of Insect Conservation, 14*, 499–510.

Merckx, T., Feber, R. E., Mclaughlan, C., Bourn, N. A. D., Parsons, M. S., Townsend, M. C., Riordan, P., & Macdonald, D. W. (2010b). Shelter benefits less mobile moth species: The field-scale effect of hedgerow trees. *Agriculture, Ecosystems and Environment, 138*, 147–151.

Merckx, T., Feber, R. E., Hoare, D. J., Parsons, M. S., Kelly, C. J., Bourn, N. A. D., & Macdonald, D. W. (2012a). Conserving threatened Lepidoptera: Towards an effective woodland management policy in landscapes under intense human land-use. *Biological Conservation, 149*, 32–39.

Merckx, T., Marini, L., Feber, R. E., & Macdonald, D. W. (2012b). Hedgerow trees and extended width field margins enhance macro-moth diversity: Implications for management. *Journal of Applied Ecology, 49*, 1396–1404.

Merckx, T., Huertas, B., Basset, Y., & Thomas, J. A. (2013). A global perspective on conserving butterflies and moths and their habitats. In D. W. Macdonald & K. J. Willis (Eds.), *Key topics in conservation biology 2* (pp. 239–257). Chichester: Wiley-Blackwell.

Monbiot, G. (2013). *Feral—Searching for enchantment on the frontiers of rewilding*. London: Penguin Books.

New, T. R. (2009). *Insect species conservation*. Cambridge: Cambridge University Press.

Öckinger, E., Eriksson, A. K., & Smith, H. G. (2006). Effects of grassland abandonment, restoration and management on butterflies and vascular plants. *Biological Conservation, 133*, 291–300.

Pereira, H. M., Daily, G. C., & Roughgarden, J. (2004). A framework for assessing the relative vulnerability of species to land-use change. *Ecological Applications, 14*, 730–742.

Pereira, H. M., Navarro, L. M., & Martins, I. S. (2012). Global biodiversity change: The bad, the good, and the unknown. *Annual Review of Environment and Resources, 37*, 25–50.

Proenca, V., & Pereira, H. M. (2013). Species-area models to assess biodiversity change in multi-habitat landscapes: The importance of species habitat affinity. *Basic and Applied Ecology, 14*, 102–114.

Proença, V., Pereira, H. M., & Vicente, L. (2010). Resistance to wildfire and early regeneration in natural broadleaved forest and pine plantation. *Acta Oecologica, 36,* 626–633.

Queiroz, C., Proença, V., Merckx, T., Guilherme, J., Ceauşu, S., Pereira, H. M., & Lindborg, R. (submitted). Habitat, scale and species richness affect functional diversity of plants, moths and birds. *Ecography.*

Rodrigues, P. (2010) Landscape changes in Castro Laboreiro: From farmland abandonment to forest regeneration. MSc Thesis, Faculdade de Ciências da Universidade de Lisboa, Portugal.

Sánchez-Zapata, J. A., & Calvo, J. F. (1999). Raptor distribution in relation to landscape composition in semi-arid Mediterranean habitats. *Journal of Applied Ecology, 36,* 254–262.

Scheper, J., Holzschuh, A., Kuussaari, M., Potts, S. G., Rundlöf, M., Smith, H. G., & Kleijn, D. (2013). Environmental factors driving the effectiveness of European agri-environmental measures in mitigating pollinator loss—a meta-analysis. *Ecology Letters, 16,* 912–920.

Sierens, T., & Van de Kerckhove, O. (2014). 32 years of moth inventories between Bruges and Ghent [In Dutch]. *Natuur.Focus 13,* 66–71.

Skórka, P., Settele, J., & Woyciechowski, M. (2007). Effects of management cessation on grassland butterflies in southern Poland. *Agriculture, Ecosystems and Environment, 121,* 319–324.

Slade, E. M., Merckx, T., Riutta, T., Bebber, D. P., Redhead, D., Riordan, P., & Macdonald, D. W. (2013). Life-history traits and landscape characteristics predict macro-moth responses to forest fragmentation. *Ecology, 94,* 1519–1530.

Steffan-Dewenter, I., Münzenberg, U., Bürger, C., Thies, C., & Tscharntke, T. (2002). Scale-dependent effects of landscape context on three pollinator guilds. *Ecology, 83,* 1421–1432.

Thomas, J. A. (2005). Monitoring change in the abundance and distribution of insects using butterflies and other indicator groups. *Philosophical Transactions of the Royal Society B: Biological Sciences, 360,* 339–357.

Thomas, J. A., Morris, M. G., & Hambler, C. (1994). Patterns, mechanisms and rates of extinction among invertebrates in the United Kingdom [and discussion]. *Philosophical Transactions of the Royal Society of London Series B: Biological Sciences, 344,* 47–54.

Tjørve, E. (2002). Habitat size and number in multi-habitat landscapes: A model approach based on species-area curves. *Ecography, 25,* 17–24.

Trematerra, P. (2013). *Isotrias penedana* sp. n. a new species of Lepidoptera (Tortricidae: Chlidanotinae: Polyorthini) from Portugal. *Journal of Entomological and Acarological Research, 45,* e1.

Tscharntke, T., Kleijn, A. M., Kruess, A., Steffan-Dewenter, I., & Thies, C. (2005). Landscape perspectives on agricultural intensification and biodiversity: Ecosystem service management. *Ecology Letters, 8,* 857–874.

van Geffen, K. G., van Grunsven, R. H., van Ruijven, J., Berendse, F., & Veenendaal, E. M. (2014). Artificial light at night causes diapause inhibition and sex-specific life history changes in a moth. *Ecology and Evolution, 4,* 2082–2089.

van Klink, R., van der Plas, F., van Noordwijk, C. G. E., Wallis de Vries, M. F., & Olff, H. (2014). Effects of large herbivores on grassland arthropod diversity. *Biological Reviews,* DOI: 10.1111/brv.12113.

van Langevelde, F., Ettema, J. A., Donners, M., Wallis de Vries, M. F., & Groenendijk, D. (2011). Effect of spectral composition of artificial light on the attraction of moths. *Biological Conservation, 144,* 2274–2281.

Vanreusel, W., & Van Dyck, H. (2007). When functional habitat does not match vegetation types: A resource-based approach to map butterfly habitat. *Biological Conservation, 135,* 202–211.

van Swaay, C. A. M. (2002). The importance of calcareous grasslands for butterflies in Europe. *Biological Conservation, 104,* 315–318.

van Swaay, C., Cuttelod, A., Collins, S., Maes, D., López Munguira, M., Šašić, M., Settele, J., Verovnik, R., Verstrael, T., Warren, M., Wiemers, M., & Wynhof, I. (2010). *European Red List of butterfies. IUCN and Butterfly Conservation Europe.* Luxembourg: Publications Office of the European Union.

Verdasca, M. J., Leitão, A. S., Santana, J., Porto, M., Dias, S., & Beja, P. (2012). Forest fuel management as a conservation tool for early successional species under agricultural abandonment: The case of Mediterranean butterflies. *Biological Conservation, 146,* 14–23.

Verhulst, J., Báldi, A., & Kleijn, D. (2004). Relationship between land-use intensity and species richness and abundance of birds in Hungary. *Agriculture, Ecosystems and Environment, 104,* 465–473.

Vickery, J., & Arlettaz, R. (2012). The importance of habitat heterogeneity at multiple scales for birds in European agricultural landscapes. In R. J. Fuller (Ed.), *Birds and habitat: Relationships in changing landscapes* (pp. 177–204). Cambridge University Press: UK.

Wilson, J. B., Peet, R. K., Dengler, J., & Pärtel, M. (2012). Plant species richness: The world records. *Journal of Vegetation Science, 23,* 796–802.

Young, M. (1997). *The natural history of moths.* Poyser Natural History: UK.

Chapter 7
Vegetation Restoration and Other Actions to Enhance Wildlife in European Agricultural Landscapes

José María Rey Benayas and James M. Bullock

Abstract Intensive farming practices are a major cause of destruction and degradation of natural vegetation throughout the world. However, in some regions including Europe, semi-natural vegetation and farmland systems harbour wildlife of conservation concern and other values. We propose widespread strategic revegetation—a type of restoration related to wildlife-friendly farming or land sharing with little competition for land—by planting woodland islets and hedgerows for ecological restoration in extensive agricultural landscapes. This approach allows wildlife enhancement, provision of a range of ecosystem services, maintenance of farmland production, and conservation of values linked to cultural landscapes. In contrast, vegetation restoration by land separation, namely secondary succession following farmland abandonment and tree planting, would provide all these benefits only at the landscape or regional scales as this restoration type is at the expense of field-level agricultural production. Furthermore, seed dispersal from revegetated elements favours passive restoration of nearby abandoned farmland and, consequently, rewilding. Revegetation of riparian systems and other actions that do not compete for land use such as introduction of bird perches, refuges for wildlife or creation of ponds would provide similar benefits. Revegetation of roadsides and roundabouts may support dispersal and spread of species but may function as ecological traps for wildlife. We provide a practioner's perspective related to land-sharing restoration actions in central Spain. We conclude that practical restoration projects—particularly strategic revegetation- are essential if we want to halt biodiversity loss and encourage the return of wildlife in agricultural landscapes.

Keywords Biodiversity · Farmland · Land separation · Land sharing · Seed dispersal · Strategic revegetation

J. M. Rey Benayas (✉)
Departamento de Ciencias de la Vida—UD Ecología, Universidad de Alcalá, 28871 Alcalá de Henares, Spain
e-mail: josem.rey@uah.es

J. M. Bullock
Centre for Ecology and Hydrology, Maclean Building, Benson Lane, Crowmarsh Gifford Wallingford, OX10 8BB Oxfordshire, UK
e-mail: jmbul@ceh.ac.uk

H. M. Pereira, L. M. Navarro (eds.), *Rewilding European Landscapes,*
DOI 10.1007/978-3-319-12039-3_7, © The Author(s) 2015

7.1   Introduction

A large part of environmental degradation is due to the expansion of the agricultural frontier in many parts of the world together with intensification of farming methods. For instance, Ellis and Ramankutty (2008) indicated that 14 of the World's 21 major biome types have agricultural use. Agricultural land covered 4.91 billion ha, ca. 38 % of the terrestrial surface, in 2011 (FAOSTAT 2013), to the detriment of natural vegetation cover. However, at the global scale, the amount of agricultural land has currently reached a plateau (Rey Benayas and Bullock 2012), with a redistribution of agricultural land from temperate areas towards the tropics (Foley et al. 2011; Rey Benayas and Bullock 2012). In the European Union (EU-27) 43 % of the land is under agriculture (FAOSTAT 2013), but this proportion is often nearly 100 % at more local scales such as the Castillian plains of Spain.

A powerful approach to countering the negative impacts of agricultural expansion and intensification is ecological restoration. Restoration actions are increasingly being implemented in response to the global biodiversity crisis, and are supported by agreements such as the global Convention for Biological Diversity—a major target of its strategic plan for 2020 is restoring at least 15 % of degraded ecosystems—and the EU Council's conclusions on biodiversity post-2010, e.g. "halting the loss of biodiversity and the degradation of ecosystem services in the EU by 2020, and restoring them in so far as feasible". Such policy initiatives are useful, but raise questions about our ability to manage and restore ecosystems to supply multiple ecosystem services and biodiversity (Bullock et al. 2011). For instance, there is often a trade-off between agricultural production that meets societal needs for food and fiber *vs.* other services and conservation of biodiversity (Pilgrim et al. 2010).

Recent discussions about the future of farming have contrasted "land sharing"— sometimes called "wildlife-friendly farming"- with "land separation". The former advocates the enhancement of the farmed environment, while the latter, also called "land sparing", advocates a separation of land designated for farming from that for conservation (Fischer et al. 2008; Phalan et al. 2011). Rey Benayas and Bullock (2012) argued that these approaches should not be seen as alternatives, but as representing the range of actions that can be best combined to enhance biodiversity and ecosystem services. Furthermore, considered broadly the land sharing/land separation approaches might be seen as a gradient rather than as a dichotomy as they represent actions at different spatial scales. However, when planning actions at specific locations, there is a true contrast between the land sharing and separation approaches (e.g. Phalan et al. 2011), as we will demonstrate. In this article we will first examine the complex role of agricultural systems in both delivering and harming wildlife (the so called "agriculture and conservation paradox", Rey Benayas et al. 2008). Then we focus on approaches to enhance wildlife—including rewilding—and associated ecosystem services in agricultural landscapes. On one hand, we will examine restoration actions that do not or hardly compete for land use to produce systems in which agricultural production is in partnership rather than in conflict with the enhancement of wildlife. Among the various restoration actions, we will pay particular attention to strategic revegetation by planting woodland islets and hedgerows. We will present a practitioner's perspective of implementation of such restoration actions in central Spain.

Cropland has mostly spread at the expense of forest land in Europe (Foley et al. 2005). Thus, on the other hand, we will focus on forest regrowth or passive restoration following farmland abandonment and tree plantations on cropland as examples of habitat restoration by land separation. Strategically revegetated elements and forest regrowth are linked by species dispersal processes. Thus, ecological restoration in farmland may maintain agricultural practices, promote wildlife return, and accelerate rewilding *sensu* Navarro and Pereira (see Chap. 1) in circumstances where the socio-ecological dynamics promote abandonment.

7.2   The Agriculture and Conservation Paradox

Few human activities are as paradoxical as agriculture in terms of their role for nature conservation. Agricultural activities are the major cause of negative environmental change worldwide. For instance, agriculture: is the main cause of deforestation; is the major threat to bird species; accounts for ca. 12 % of total direct global anthropogenic emissions of greenhouse gasses; and strongly impacts on soil carbon and nutrients (sources of evidence in Rey Benayas and Bullock 2012). In recent history, in addition to an increase in farmed area, farming practices in many regions have become more intensive. For example, the area of cultivated land serviced by irrigation, the major form of human water consumption and a surrogate of farmland intensification, in Europe increased from 9.2×10^6 ha in 1961 to 17.9×10^6 ha in 2011 (FAOSTAT 2013). Beyond changes in species richness, agricultural intensification has been shown to reduce the functional diversity of plant and animal communities, potentially imperilling the provisioning of ecosystem services (Flynn et al. 2009). Importantly, intensification of land use has brought remnant areas of natural or semi-natural vegetation such as steep hillsides, property boundaries and track edges into mainstream agriculture (Rey Benayas et al. 2008). Thus, agricultural expansion and intensification have greatly increased our food, fiber and biofuels supplies, but have damaged wildlife and other services.

In contrast to these negative perspectives, extensive agricultural habitats are often viewed positively in terms of nature conservation due to, for example, creation of landscape mosaics and environmental heterogeneity (Oliver et al. 2010), or because they are threatened habitats that support endangered species and cultural values (Kleijn et al. 2006). In the EU-27, 31 % of Natura 2000 sites, a network of protected areas, result from agricultural land management. Several taxa including species of birds, insects and plants, some of them endangered, depend on low-intensity farmland for their persistence (Kohler et al. 2008). Thus, common farmland birds in Europe show negative trends (−35 % since 1980) and these are today of conservation concern, whereas forest birds show positive trends due to abandonment of agricultural land and afforestation programs (European Bird Census Council 2010). Wildlife decline might affect agricultural production itself. For instance, insects that

Fig. 7.1 Sketch of a hypothetical Mediterranean agricultural landscape before (*top*) and a few years after (*bottom*) implementing strategic revegetation actions. The actions illustrated are the following: (1) introduction of woodland islets and (2) hedgerows in cropfields; (3) restoration of riparian vegetation; (4) revegetation of road sides and (5) roundabouts. Additionally, there are some (6) abandoned fields, which are indicated by *arrows*. The lack of revegetation actions in the *bottom left* quarter of the landscape illustrates the inappropriateness of such revegetation due to e.g. outstanding values linked to steppe birds. Establishment and development of vegetation following cropland abandonment is different in fields close (*blue arrows*) or away (*red arrows*) from strategic revegetation actions or natural vegetation

provide pollination and pest control services in cropland tend to be less common in more intensive landscapes (Tscharntke et al. 2005).

Agricultural intensification can have a negative impact on the values linked to traditional agriculture, but so can agricultural abandonment and, particularly, when afforestation occurs on former cropland (Rey Benayas et al. 2007). Abandonment of agricultural land has mostly occurred in developed countries in the last few decades (Rey Benayas and Bullock 2012). The European Agrarian Policy has aided afforestation in agricultural land that has resulted in the convesion of $> 10^6$ ha of former cropland into tree plantations (Directorate-General for Agriculture and Rural Development 2012). Currently, it is providing subsidies to afforest land after vineyard extirpation in Spain, for instance, an action that is being criticized by conservationists due to negative impacts on wildlife and other values (Rey Benayas and Bullock 2012). It seems that agriculture, woodland, and biological conservation are in a permanent and irreconcilable conflict, *the agriculture and conservation paradox* (Rey Benayas et al. 2008).

7.3   Designing Restoration on Agricultural Land by Strategic Revegetation

The agriculture and conservation paradox creates a dilemma in projects that involve restoring non-agricultural habitats such as woodland on agricultural land, which can only be resolved by considering the relative values of biodiversity and ecosystem services associated with woodland *vs.* agricultural ecosystems (Rey Benayas et al. 2008). The reconstruction of vegetation in a landscape ("where and when to revegetate?") is an issue that has become a research priority (Thompson et al. 2009). In the context of this article, we consider as strategic revegetation, highly specific planting (and sometimes seeding) actions that are characterized as occupying a tiny fraction of the agricultural landscape (Fig. 7.1). They are intended to enhance wildlife and particular services such as habitat provision and seed dispersal. The effects on wildlife and ecosystem services will usually depend on how much land is affected by these actions.

Strategic Revegetation in Farmed Fields

In actively farmed fields, these actions can include planting woodland islets, hedgerows and isolated trees (Fig. 7.1). They have the potential to enhance wildlife, agricultural production, and other services at the field and landscape scales since they hardly compete for farmland use (Rey Benayas and Bullock 2012), and can be considered a form of rewilding *per se*. Rey Benayas et al. (2008) suggested a new

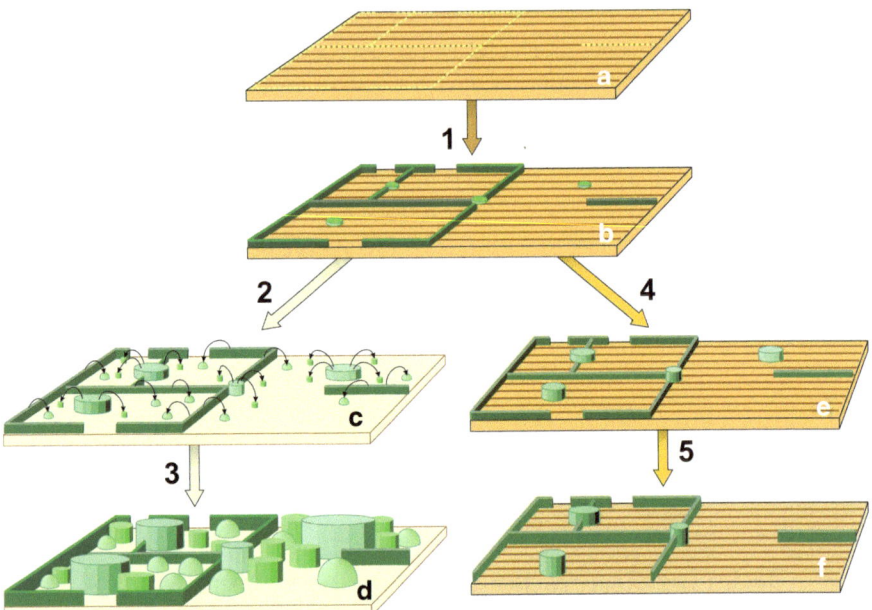

Fig. 7.2 A schematic diagram of the "woodland islet and hedgerow" model proposed in this article, based on the "woodland islets" model developed by Rey Benayas et al. (2008). A denuded agricultural landscape (**a**) is planted with a few (here four) small (e.g. 100 m²) woodland islets and hedgerows (**b**) Targeted management of the islets and hedgerows allows the trees to establish, grow and reach sexual maturity rapidly. If the cropland is then abandoned the islets and hedgerows can expand and export seeds (and other organisms established in them) to the surrounding land (**c**) The islets and hedgerows eventually coalesce to form closed woodland (**d**) Alternatively the surrounding land remains in same or other uses (**e**) while the islets and hedgerows remain as small patches of the native woodland community as the trees grow taller. Some islets and hedgerow fragments may disappear through stochastic events (**f**)

concept for designing restoration of forest ecosystems on agricultural land, which uses small-scale active restoration as a driver for passive restoration over much larger areas. Establishment of "woodland islets" is an approach to designing restoration of woodlands in extensive agricultural landscapes where no remnants of native natural vegetation exist. It involves planting a number of small, densely-planted, and sparse blocks of native shrubs and trees within agricultural land that together occupy a tiny fraction of the area (< 1 %) of target land to be restored (Fig. 7.2). This approach, later called "applied nucleation" by Corbin and Holl (2012), allows direction of secondary succession by establishing small colonisation *foci*, while using a fraction of the resources required for large-scale reforestation. Woodland patches provide sources of seed and dispersing animals that can colonize adjacent habitats. If the surrounding land is abandoned, colonists from the islets could accelerate woodland development because dispersal of many woodland organisms will

Table 7.1 Examples of seed dispersal distances for European forest and scrub species, with information on whether dispersal is biased into open or forest habitats

Species	Dispersal vector(s)	Maximum reported dispersal distance (m)	Biased dispersal among different habitats?	Reference
Fraxinus excelsior	Wind	90	None	Stoyan and Wagner (2001)
Picea abies	Wind	150	None	Dovciak et al. (2008)
Pinus sylvestris, P. nigra	Wind	100	None	Debain et al. (2007)
Corylus avellana	Rodents	>70	Grassland > woodland	Laborde and Thompson (2009)
Corylus avellana, Quercus petraea	Jays (Garrulus glandarius), mice	>200	Jays: short grass> tall grass/woodland Mice: grassland> woodland	Kollmann and Schill (1996)
Fagus sylvatica	Jays, rodents	3000	Pine forest > grassland	Kunstler et al. (2007)
Prunus mahaleb	Birds, mammals	>1500	Large birds and mammals: open> scrub Small birds: scrub> open	Jordano et al. (2007)
Quercus ilex, Q. suber	Jays	550	Abandoned field/forest track > managed field/scrub	Pons and Pausas (2007)
Quercus robur, Q. petraea	Mice, voles	>137	Heath > woodland/ grassland	Jensen and Nielsen (1986)

continue over many years (Fig. 7.1). The landscape emphasis on a planned planting of islets maximises benefits to wildlife and the potential of allowing the islets to trigger larger-scale reforestation if the surrounding land is abandoned, which can lead to rewilding (see Chap. 1). The islets should be planted with a variety of native shrub and tree species including those identified as nurse species to take advantage of facilitation processes (Cuesta et al. 2010).

Vegetation dynamics in complex landscapes depend on interactions among environmental heterogeneity, disturbance, habitat fragmentation, and seed dispersal processes. Ozinga et al. (2009) concluded that the 'colonization deficit' of plant species due to a degraded dispersal infrastructure is as important in explaining plant diversity losses as habitat quality, and called for new measures to restore the dispersal infrastructure across entire regions. Estimates of dispersal distances for vertebrate-dispersed shrubs and trees (Table 7.1) suggest that the introduction of woodland islets planted about one km apart in a deforested agricultural landscape could allow colonisation over the whole landscape (Rey Benayas and Bullock 2012). Spread from these islets would be facilitated in the cases where animals disperse seeds preferentially into open habitats, whilst avoiding dense scrub or forest (Table 7.1). However, the potential for colonisation from such *foci* will be more limited for wind-dispersed trees and shrubs, which seem to disperse shorter distances, and in those cases where animals disperse seeds preferentially into wooded habitats (Table 7.1). It is possible however to direct dispersal artificially into open habitats; for example by erecting structures such as perches or hedges which attract birds and/or which act as a physical barrier to wind-dispersed seeds (Bullock and Moy 2004).

The woodland islets approach maintains flexibility of land use, which is critical in agricultural landscapes where land use is subject to a number of fluctuating social, environmental, policy and economic drivers (e.g. Romero-Calcerrada and Perry 2004; Rounsevell et al 2005). It provides a means of reconciling competition for land use among agriculture, conservation and woodland restoration at the landscape scale. This could increase the economic feasibility of large-scale restoration projects and facilitate the involvement of local human communities in the restoration process. The woodland islets idea has similarities to other approaches involving planting small areas of trees on farms, such as tree clumps, woodlots, hedges, living fences, or shelterbelts and agro-forestry systems. Particularly, the revegetation of property boundaries, field margins and track edges in farmland to create living fences (Barnes and Williamson 2006; see Chap. 6) has the same function in triggering passive revegetation as woodland islets (Forget et al. 2013); thus, the "woodland islets" concept could be expanded to the "woodland islets and hedgerows" concept (Fig. 7.2). Planting isolated trees may also provide a disproportionate positive value for wildlife and potential for seed dispersal (DeMars et al. 2010; Fischer et al. 2010).

Besides providing a dispersal infrastructure, woodland islets and hedgerows provide habitat or enhance the farmed environment for wildlife. These benefits have been well documented for plant species (e.g. Poggio et al. 2010) and small animals such as insects (e.g. Noordijk et al. 2010), but they are also critical for medium-sized and even large mammals. For instance, Pereira and Rodríguez (2010) documented the value of hedgerows and narrow strips of riparian forest for the Egyptian mongoose *Herpestes ichneumon* and the common genet *Genetta genetta*. They found that mongooses and genets strongly selected linear woody vegetation, and that open fields, *dehesa* and olive groves were avoided. Similarly, Blanco and Cortés (2007) demonstrated that hedgerows and small woodland patches were important for wolf *Canis lupus* –an iconic species for the rewilding concept- living in agricultural habitats in northern Spain.

Other Options for Strategic Revegetation in Agricultural Landscapes

While we concentrate on woodlands here, the islets approach to restoring a dispersal infrastructure could be used for other (semi-)natural habitats such as species-rich grasslands, scrub, or heathland (e.g. Hooftman and Bullock 2012). Other strategic revegetation actions in agricultural landscapes but unrelated to the farmed environment itself could, for example, target road verges and roundabouts, and riparian systems (Fig. 7.1). These would provide similar benefits in terms of dispersal infrastructure as those explained above for woodland islets and hedgerows. The revegetation of roadsides and roundabouts may bring additional benefits such as slope stabilization and aesthetic value. However, these revegated artificial elements may also function as "ecological traps" that put at risk the wildlife attracted by them through increasing traffic collisions (Fahrig and Rytwinski 2009). Additionally, they may also decrease visibility for drivers and be dangerous obstacles in case of crashes, so safety considerations should also be considered before revegetation is decided. Thus, they should be carefully planned and supported by the planners and contractors.

Riparian systems often support the only natural or semi-natural communities at the local level in agricultural landscapes, but frequently this vegetation has been extirpated or highly degraded, and the riverside has been ploughed. It has been shown that riparian vegetation is critical for wildlife conservation (Forget et al. 2013) and provision of ecosystem services such as water regulation and purification; thus, we suggest that strategic revegetation of creeks and rivers with native species should be considered a priority in agricultural landscapes and enforced by competent public administrations.

7.4   Restoring or Creating Other Specific Elements to Benefit Wildlife and Particular Services

Besides strategic revegetation, there are other actions that benefit wildlife and provide particular services in farmland, which do not or hardly compete with agricultural land use. In general, all these actions were labeled "farmed field manicure" by Rey Benayas (2012) and, again, they can be considered as a form of rewilding. These include: (1) creation of pollinator-friendly areas using plant enrichment; (2) introduction of beetle banks, stone walls, stone mounds and other strategic refuges for fauna; (3) introduction of perches and nest-boxes for birds (see example below); (4) introduction or restoration of small ponds and (5) drinking troughs; and (6) reconstruction of rural architecture aiming at restoring cultural services.

GREFA's project for enhancement of birds of prey for rodent control (http://www. grefa.org/proyectosgrefa/38-proyectos/servivios-ambientales/control-biologico-del-topillo-campesino/76-control-biologico-del-topillo-campesino) is an outstanding example of this type of wildlife-friendly farming. This project was motivated by periodic field vole *Microtus arvalis* outbreaks, which are often controlled using poisons that may damage wildlife and game. Common kestrel *Falco tinnunculus* and barn owl *Tyto alba* are rodent predators that have declining populations for a number of reasons, including lack of sites for nesting in open landscapes. Thus, more nesting sites should increase the populations of these two species and contribute to place their populations at the carrying capacities. To achieve this goal, three

2000-ha agricultural landscapes in central Spain were seeded with nest boxes, 100 for common kestrel and 100 for barn owl in each landscape. For common kestrel, we calculate that rodent consumption per occupied nest box is ca. 186 kg year^{-1}. As average occupancy in the three landscapes was 27 % between 2009 and 2012, total rodent consumption by this species is calculated in ca. 5 t year^{-1} per landscape for those years. Total rodent consumption could be as high as ca. 46 t year^{-1} if full nest occupancy by both species was attained, a figure that is expected to contribute to both rodent damage control and the maintenance of these birds of prey.

7.5   A Practitioner's Perspective

The International Foundation for Ecosystem Restoration (FIRE, www.fundacionfire.org) aims at translating academic knowledge to ecosystem restoration in the real world, an example of translational ecology. It provides leadership in implementing restoration actions in farmland habitat and farmland stewardship in Spain by means of its "Fields for Life Initiative", which targets reconciliation of agricultural production and wildlife enhancement based on sound, targeted research. Since 2008, this initiative has revegetated 6.5 km of hedgerows and three woodland islets of different size with ca. 12,600 seedlings of 27 native species, introduced nine artificial ponds and several hundreds of artificial nests for insectivorous birds and 121 for birds of prey, and has completed 12 signed stewardship agreements with land owners, among other achivements, including the participation of hundreds of volunteers in such actions, mostly in central Spain. The total area involved in this project is 3358 ha so far.

During these years, we have learnt that, in the first instance, farmers are reluctant to implement the suggested revegetation projects. First, farmers do not understand or forsee the benefits for agricultural production and, simultaneously, they perceive risks for crops. For instance, they believe that a new planted hedgerow is a reservoir that will spread crop pests rather than habitat for natural enemies of such pests or pollinators. They also think that the role of hedgerows as windbrakes that reduce soil erosion and dessication and crop abrasion is irrelevant for crop production. The second major reason has to do with their aesthetic appraisal of crop fields. According to their perception, crop fields must be "clean", i.e. with nothing other than the cultivated plants, and most often farmers that have "untidy" crop fields are criticized in their local communities. And third, generally, individual farmers react to the private use-value of biodiversity and ecosystem services assigned in the marketplace and thus typically ignore the 'external' benefits of conservation that accrue to wider society (Jackson et al. 2007). To overcome this reluctance, we recommend efforts to educate and show farmers that strategic revegetation and other actions benefit wildlife and wildlife-based ecosystem services that may enhance or be neutral for crop production (see also De Snoo et al. 2013). There is a need to address the inertia in farmers' perceptions, and the EU Common Agricultural Policy should provide specific resources to target this social objective for agricultural landowners beyond simple financial support such as the agri-environment schemes.

In contrast to such negative perceptions, and in addition to the obvious positive opinions of naturalists and conservationists, our projects are best valued and encouraged by hunters. They understand that planting woodland islets and hedgerows, the creation of ponds and other restoration actions are very beneficial to game, including birds such as the Red-legged partridge *Alectoris rufa*, rabbits and hares. Enhancement of game production and its associated economical benefits for local communities is an incentive for strategic restoration of agricultural fields.

The reported reluctance of farmers may be overcome; for that, we have learnt that the form of first contact with farmers is very important. The farmers need quite a lot of time to understand the possible advantages and, in the worst case, the overall non-harmful character of restoration actions in their properties. As a stewardship agreement is voluntary, it is necessary to have a continuing but 'light touch' contact with farmers to persuade them to undertake necessary actions. Once a farmer agrees to implement restoration actions on his land, other farmers in a local community often agree too. In, unfortunately, few cases we have found landowners that are rapidly persuaded to adopt restoration actions, but that are almost never willing to pay any of the cost. Thus, FIRE seeks public and private financial support for its projects on the basis of their demonstration value. In short, key issues for large-scale ecological restoration on agricultural land are financial support and education to promote farmer and public awareness and training (Rey Benayas and Bullock 2012; De Snoo et al. 2013). Land owners must be specifically rewarded for restoration actions on their properties. To reward the total or social value, tax deductions for land owners who restore agricultural land and donations to not-for-profit organizations that run restoration projects, payment for environmental services, and direct financing measures related to restoration activities should be implemented widely.

7.6   Forest Restoration by Land Separation

Rewilding by setting aside farmland to restore or create non-farmed habitat rarely happens –except in the case of farmland abandonment- as farmers tend to use and expand into all available land since this is usually the most profitable choice in terms of direct use value (TEEB 2010). This approach to rewilding competes with land for agricultural production at the field scale. Nevertheless, rewilding and agricultural production can coexist at the regional scale by a combination of habitat restoration and creation and maintenance of productive land for rewilding. Thus, rewilding *sensu* Navarro and Pereira (Chap. 1) might be considered more as land separation at the local scale, but it could also be seen a land sharing option at larger scales. Two major contrasting approaches for large-scale woodland or forest restoration in agricultural landscapes are: (1) passive restoration through secondary succession or forest regrowth following abandonment of agricultural land, e.g. cropland

and pastures where extensive livestock farming has been removed; and (2) active restoration through addition of desired plant species. Forest regrowth and tree plantations on cropland enhance species that are characteristic of shrubland and forest environments, but are detrimental to species that are characteristic of open farmland environments and to agricultural production (Rey Benayas and Bullock 2012).

Passive restoration is cheap (although it may include opportunity costs) and leads to a local vegetation type. It is generally fast in productive environments, but slow in low productivity environments, as woody vegetation establishment is limited (Rey Benayas et al. 2008). The restoration capacity of woody ecosystems depends on the magnitude and duration of ecosystem modification, i.e., the "agricultural legacy" (Dwyer et al. 2010). A key bottleneck that hinders revegetation in large, continuous agricultural landscapes is the lack of propagules due to absence of parent trees and shrubs (García et al. 2010), which might be overcome by strategic revegetation as explained above. Passive restoration can be seen as a rewilding process (see Chap. 1), and it is of particular importance for large carnivores and herbivores such as the Brown bear *Ursus arctos* and the European bison *Bison bonasus*. The reintroduction of these species, whose habitat is expanding due to land abandonment, is often the subject of heated debates. The International Union for Conservation of Nature (2012) has recently published a document related to reintroductions and other translocations that is much more flexible than its previous 1998 Guide for such actions.

Active forest restoration basically comprises the planting of trees and shrubs (Stanturf et al. 2014). It is needed, for example, when abandoned land suffers continuing degradation, local vegetation cover cannot be recovered and secondary succession has to be accelerated. Indeed, one criticism of the passive rewilding approach is that the establishment of forest and other natural habitats in degraded landscapes may be impossible without more active interventions (Hodder and Bullock 2010). There are differences in the wildlife and ecosystem services provided by passive *vs.* active restoration, and there is much debate about the ecological benefits of tree plantations. For instance, Bremer and Farley (2010) found that plantations are most likely to contribute to biodiversity when established on degraded lands rather than replacing natural ecosystems, and when indigenous tree species are used rather than exotic species. Similarly, a meta-analysis of faunal and floral species richness and abundance in timber plantations and pasture lands on 36 sites across the world concluded that plantations support higher species richness or abundance than pasture land only for particular taxonomic groups (i.e. herpetofauna), or specific landscape features (i.e. absence of remnant vegetation within pasture) (Felton et al. 2010). Cropland afforestations in southern Europe, which are mostly based on coniferous species, may cause severe damage to populations of open habitat species, especially birds, by replacing high quality habitat and increasing risk of predation (Reino et al. 2010). Further, these planted forests have been shown to be suitable habitat for generalist forest birds but not for specialist forest birds (Sánchez-Oliver et al. 2014), whereas secondary succession shrubland and woodland favour bird species that are of conservation concern in Europe (Rey Benayas et al. 2010).

7.7   Conclusions

We conclude that, although agriculture is a major cause of environmental degradation, ecological restoration on agricultural land offers opportunities to reconcile agricultural production with enhancement of wildlife and ecosystem services other than production. Strategic revegetation by land sharing through environmentally-friendly farming, namely planting woodland islets, hedgerows and isolated trees, has the potential to enhance agricultural production, other ecosystem services and wildlife at both the farmed field and landscape scale. Importantly, strategic revegetation has the potential to trigger larger-scale reforestation if the surrounding land is abandoned (rewilding). However, vegetation restoration by land separation, namely secondary succession following farmland abandonment and tree plantations, would provide these triple benefits only at the landscape or regional scales as this restoration type is at the expense of field-level agricultural production. At the landscape level, restoration of riparian vegetation is a priority whereas strategic revegation of road sides and roundabouts should be carefully planned to avoid risks for wildlife and drivers. Beyond scientific and technical research, an increase in such restoration projects is needed if we want to halt environmental degradation and biodiversity loss and meet the CBD and UE goals (see Chap. 11). We propose widespread expansion of highly specific actions to benefit wildlife and specific ecosystem services, particularly habitat provision and seed dispersion for triggering passive restoration after land abandonment leading to rewilding. Financial support, public awareness, education and training, particularly of farmers, are necessary to accomplish such objectives.

Acknowledgments Figures 7.1 and 7.2 were produced by Alejandra Toledo and by Luis Monje, respectively, following the authors' directions. Henrique Pereira and Laetitia Navarro provided useful comments on a previous version of this manuscript. Projects from the Spanish Ministry of Science and Education (CGL2010-18312), the Government of Madrid (S2009AMB-1783, REMEDINAL-2) and NERC (Wessex-BESS) are currently providing financial support for this body of research.

References

Barnes, G., & Williamson, T. (2006). *Hedgerow history: Ecology, history and landscape character*. Macclesfield: Windgather Press.
Blanco, J. C., & Cortés, Y. (2007). Dispersal patterns, social structure and mortality of wolves living in agricultural habitats in Spain. *Journal of Zoology, 273,* 114–124.
Bremer, L. L., & Farley, K. A. (2010). Does plantation forestry restore biodiversity or create green deserts? A synthesis of the effects of land-use transitions on plant species richness. *Biodiversity Conservation, 19,* 3893–3915.

Bullock, J. M., & Moy, I. L. (2004). Plants as seed traps: Inter-specific interference with dispersal. *Acta Oecologica, 25,* 35–41.

Bullock, J. M., Aronson, J., Newton, A., C., Pywell, R. F., & Rey-Benayas, J. M. (2011). Restoration of ecosystem services and biodiversity: Conflicts and opportunities. *Trends in Ecology and Evolution, 26,* 541–549.

Corbin, J. D., & Holl, K. D. (2012). Applied nucleation as a forest restoration strategy. *Forest Ecology and Management, 265,* 37–46.

Cuesta, B., Villar-Salvador, P., Puértolas, J., Rey Benayas, J. M., & Michalet, R. (2010). Facilitation of *Quercus ilex* in Mediterranean shrubland is explained by both direct and indirect interactions mediated by herbs. *Journal of Ecology, 98,* 687–696.

de Snoo, G. R., Herzon, I., Staats, H., Burton, R. J. F., Schindler, S., van Dijk, J., Lokhorst, A. M., Bullock, J. M., Lobley, M., Wrbka, T., Schwarz, G., & Musters, C. J. M. (2013). Toward effective nature conservation on farmland: Making farmers matter. *Conservation Letters, 6,* 66–72.

Debain, S., Chadaeuf, J., Curt, T., Kunstler, G., & Lepart, J. (2007). Comparing effective dispersal in expanding population of *Pinus sylvestris* and *Pinus nigra* in calcareous grassland. *Canadian Journal of Forest Research, 37,* 705–718.

DeMars, C. A., Rosenberg, D. K., & Fontaine, J. B. (2010). Multi-scale factors affecting bird use of isolated remnant oak trees in agro-ecosystems. *Biological Conservation, 143,* 1485–1492.

Directorate-General for Agriculture and Rural Development. (2012). Rural development in the European Union. Statistical and Economic Information. Report 2012. Brussels.

Dovciak, M., Hrivnak, R., Ujhazy, K., & Gomory, D. (2008). Seed rain and environmental controls on invasion of *Picea abies* into grassland. *Plant Ecology, 194,* 135–148.

Dwyer, J. M., Fensham, R. J., & Buckley, Y. M. (2010). Agricultural legacy, climate, and soil influence the restoration and carbon potential of woody regrowth in Australia. *Ecological Applications, 20,* 1838–1850.

Ellis, E. C., & Ramankutty, N. (2008). Putting people in the map: Anthropogenic biomes of the world. *Frontiers in Ecology and the Environment, 6,* 439–447.

European Bird Census Council. (2010). Pan-European common bird monitoring scheme (on line). http://www.ebcc.info/. Accessed 28 Feb 2012.

Fahrig, L., & Rytwinski, T. (2009). Effects of roads on animal abundance: An empirical review and synthesis. *Ecology and Society, 14*(1), 21.

FAOSTAT. http://faostat3.fao.org/home/index.html#DOWNLOAD. Accessed 9 April 2013.

Felton, A., Knight, E., Wood, J., Zammit, C., & Lindenmayer, D. (2010). A meta-analysis of fauna and flora species richness and abundance in plantations and pasture lands. *Biological Conservation, 143,* 545–554.

Fischer, J., Brosi, B., Daily, G. C., Ehrlich, P. R., Goldman, R., Goldstein, J., Lindenmayer, D. B., Manning, A. D., Mooney, H. A., Pejchar, L., Ranganathan, J., & Tallis, H. (2008). Should agricultural policies encourage land sparing or wildlife-friendly farming? *Frontiers in Ecology and the Environment, 6,* 382–387.

Fischer, J., Stott, J., & Law, B. S. (2010). The disproportionate value of scattered trees. *Biological Conservations, 143,* 1564–1567.

Flynn, D. F. B., Gogol-Prokurat, M., Nogeire, T., Molinari, N., Richers, B. T., Lin, B. B., Simpson, N., Mayfield, M. M., & DeClerck, F. (2009). Loss of functional diversity under land use intensification across multiple taxa. *Ecology Letters, 12,* 22–33.

Foley, J. A., DeFries, R., Asner, G. P., Barford, C., Bonan, G., Carpenter, S. R., Chapin, F. S., Coe, M. T., Daily, G. C., Gibbs, H. K., Helkowski, J. H., Holloway, T., Howard, E. A., Kucharik, C. J., Monfreda, C., Patz, J. A., Prentice, I. C., Ramankutty, N., & Snyder, P. K. (2005). Global consequences of land use. *Science, 309,* 570–574.

Foley, J. A., Ramankutty, N., Brauman, K. A., Cassidy, E. S., Gerber, J. S., Johnston, M., Mueller, N. D., O'Connell, C., Ray, D. K., West, P. C., Balzer, C. B., Bennett, L. M., Carpenter, S. R., Hill, J., Monfreda, C., Polasky, S., Rockstrom, J., Sheehan, J., Siebert, S., Tilman, D., & Zaks, D. P. M. (2011). Solutions for a cultivated planet. *Nature, 478,* 337–342.

Forget, G., Carreau, C., Le Coeur, D., & Bernez, I. (2013). Ecological restoration of headwaters in a rural landscape (Normandy, France): A passive approach taking hedge networks into account for riparian tree recruitment. *Restoration Ecology, 21,* 96–104.

García, D., Zamora, R., & Amico, G. (2010). Birds as suppliers of seed dispersal in temperate ecosystems: Conservation guidelines from real-world landscapes. *Conservation Biology, 24,* 1070–1079.

Hodder, K. H., & Bullock, J. M. (2010). Nature without nurture. In M. Hall (Ed.), *Restoration and history: The search for a usable environmental Past* (pp. 223–235). Abingdon: Routledge.

Hooftman, D. A. P., & Bullock, J. M. (2012). Mapping to inform conservation: A case study of changes in semi-natural habitats and their connectivity over 70 years. *Biological Conservation, 145,* 30–38.

International Union for Conservation of Nature. (2012). IUCN Guidelines for reintroductions and other conservation translocations. http://www.issg.org/pdf/publications/Translocation-Guidelines-2012.pdf. Accessed 4 April 2013.

Jackson, L. E., Pascual, U., & Hodgkin, T. (2007). Utilizing and conserving agrobiodiversity in agricultural landscapes. *Agriculture, Ecosystems and the Environment, 121,* 196–210.

Jensen, T. S., & Nielsen, O. F. (1986). Rodents as seed dispersers in a heath oak wood succession. *Oecologia, 70,* 214–221.

Jordano, P., Garcia, C., Godoy, J. A., & Garcia-Castano, J. L. (2007). Differential contribution of frugivores to complex seed dispersal patterns. *Proceedings of the National Academy of Sciences, 104,* 3278–3282.

Kleijn, D., Baquero, R. A., Clough, Y., Díaz, M., Esteban, J., Fernández, F., Gabriel, D., Herzog, F., Holzschuh, A., Jöhl, R., Knop, E., Kruess, A., Marshall, E. J., Steffan-Dewenter, I., Tscharntke, T., Verhulst, J., West, T. M., & Yela, J. L. (2006). Mixed biodiversity benefits of agro-environment schemes in five European countries. *Ecology Letters, 9,* 243–254.

Kohler, F., Vandenberghe, C., Imstepf, R., & Gillet, F. (2008). Restoration of threatened arable weed communities in abandoned mountainous crop fields. *Restoration Ecology, 19,* 62–69.

Kollmann, J., & Schill, H. P. (1996). Spatial patterns of dispersal, seed predation and germination during colonization of abandoned grassland by *Quercus petraea* and *Corylus avellana*. *Vegetatio, 125,* 193–205.

Kunstler, G., Thuiller, W., Curt, T., Bouchaud, M., Jouvie, R., Deruette, F., & Lepart, J. (2007). *Fagus sylvatica* L. recruitment across a fragmented Mediterranean Landscape, importance of long distance effective dispersal, abiotic conditions and biotic interactions. *Diversity and Distributions, 13,* 799–807.

Laborde, J., & Thompson, K. (2009). Post-dispersal fate of hazel (*Corylus avellana*) nuts and consequences for the management and conservation of scrub-grassland mosaics. *Biological Conservation, 142,* 974–981.

Noordijk, J., Musters, C. J. M., van Dijk, J., & de Snoo GR. (2010). Invertebrates in field margins: taxonomic group diversity and functional group abundance in relation to age. *Biodiversity Conservation, 19,* 3255–3268.

Oliver, T., Roy, D. B., Hill, J. K., Brereton, T., & Thomas, C. D. (2010). Heterogeneous landscapes promote population stability. *Ecology Letters, 13,* 473–484.

Ozinga, W. A., Romermann, C., Bekker, R. M., Prinzing, A., Tamis, W. L. M., Schaminee, J. H. J., Hennekens, S. M., Thompson, K., Poschlod, P., Kleyer, M., Bakker, J. P., & van Groenendael, J. M. (2009). Dispersal failure contributes to plant losses in NW Europe. *Ecology Letters, 12,* 66–74.

Pereira, M., & Rodríguez, A. (2010). Conservation value of linear woody remnants for two forest carnivores in a Mediterranean agricultural landscape. *Journal of Applied Ecology, 47,* 611–620.

Phalan, B., Onial, M., Balmford, A., & Green, R. E. (2011). Reconciling food production and biodiversity conservation: Land sharing and land sparing compared. *Science, 333,* 1289–1291.

Pilgrim, E. S., Macleod, C. J. A., Blackwell, M. S. A., Bol, R., Hogan, D. V., Chadwick, D. R., Cardenas, L., Misselbrook, T. H., Haygarth, P. M., Brazier, R. E., Hobbs, P., Hodgson, C., Jarvis, S., Dungait, J., Murray, P. J., Firbank, L. G., & Donald, L. S. (2010). Interactions among agricultural production and other ecosystem services delivered from European temperate grassland systems. *Advances in Agronomy, 109,* 117–154.

Poggio, S. L., Chaneton, E. J., & Ghersa, C. M. (2010). Landscape complexity differentially affects alpha, beta, and gamma diversity of plants occurring in fencerows and crop fields. *Biological Conservation, 143*, 2477–2486.

Pons, J., & Pausas, J. G. (2007). Acorn dispersal estimated by radio-tracking. *Oecologia, 153,* 903–911.

Reino, L., Porto, M., Morgado, R., Carvalho, F., Mira, A., & Beja, P. (2010). Does afforestation increase bird nest predation risk in surrounding farmland? *Forest Ecology and Management, 260,* 1359–1366.

Rey Benayas, J. M. (2012). Restauración de campos agrícolas sin competir por el uso de la tierra para aumentar su biodiversidad y servicios ecosistémicos. Investigación Ambiental. *Ciencia y Política Pública, 4,* 101–110.

Rey Benayas, J. M., & Bullock, J. M. (2012). Restoration of biodiversity and ecosystem services on agricultural land. *Ecosystems, 15,* 883–889.

Rey Benayas, J. M., Martins, A., Nicolau, J. M., & Schulz, J. (2007). Abandonment of agricultural land: an overview of drivrs and consequences. *Perspectives in Agriculture, Veterinary Sciences, Nutrition and Natural Resources, 2*(057), 1–14.

Rey Benayas, J. M., Bullock, J. M., & Newton, A. C. (2008). Creating woodland islets to reconcile ecological restoration, conservation, and agricultural land use. *Frontiers in Ecology and the Environment, 6,* 329–336.

Rey Benayas, J. M., Galván, I., & Carrascal, L. M. (2010). Differential effects of vegetation restoration in Mediterranean abandoned cropland by secondary succession and pine plantations on bird assemblages. *Forest Ecology and Management, 260,* 87–95.

Romero-Calcerrada, R., & Perry, G. L. W. (2004). The role of land abandonment in landscape dynamics in the SPA 'Encinares del rio Alberche y Cofio, Central Spain, 1984–1999. *Landscape and Urban Planning, 66,* 217–232.

Rounsevell, M. D. A., Ewert, F., Reginster, I., Leemans, R., & Carter, T. R. (2005). Future scenarios of European agricultural land use II. Projecting changes in cropland and grassland. *Agriculture, Ecosystems & Environment, 107,* 117–135.

Sánchez-Oliver, J., Rey Benayas, J. M., & Carrascal, L. M. (2014). Differential effects of local habitat and landscape characteristics on bird communities in Mediterranean afforestations motivated by the EU Common Agrarian Policy. *European Journal of Wildlife Research, 60,* 135–143.

Stanturf, J. A., Palik, B. J., & Dumroese, R. K. (2014). Contemporary forest restoration: A review emphasizing function. *Forest Ecology and Management, 331,* 292–323.

Stoyan, D., & Wagner, S. (2001). Estimating the fruit dispersion of anemochorous forest trees. *Ecological Modeling, 45,* 35–47.

TEEB. (2010). The economics of ecosystems and biodiversity: Mainstreaming the economics of nature: A synthesis of the approach, conclusions and recommendations of TEEB. http://www.teebweb.org/. Accessed 10 Sep 2013.

Thompson, J. R., Moilanen, A., Vesk, P. A., Bennett, A. F., & Mac Nally, R. (2009). Where and when to revegetate?: A quantitative method for scheduling landscape reconstruction. *Ecological Applications, 19,* 817–827.

Tscharntke, T., Klein, A. M., Kruess, A., Steffan-Dewenter, I., & Thies, C. (2005). Landscape perspectives on agricultural intensification and biodiversity-ecosystem service management. *Ecology Letters, 8,* 857–874.

Chapter 8
Maintaining Disturbance-Dependent Habitats

Laetitia M. Navarro, Vânia Proença, Jed O. Kaplan and Henrique M. Pereira

Abstract Natural disturbances, or the lack thereof, contributed to shape Earth's land-scapes and maintain its diversity of ecosystems. In particular, natural fire dynamics and herbivory by wild megafauna played an essential role in defining European land-scapes in pre-agricultural times. The advent of agriculture and the development of complex societies exacerbated the decline of European megafauna, leading to local and global extinctions of many species, and substantial alterations of fire regimes. Those natural phenomena were over time gradually and steadily replaced by anthropogenic disturbances. Yet, for the first time since the Black Death epidemic, agricultural land-use is decreasing in Europe. Less productive marginal areas have been progressively abandoned as crop and livestock production has become concentrated on the most fertile and easier to cultivate land. With little or no substitute for the anthropogenic disturbances associated with these abandoned agricultural practices, there is growing concern that disturbance-dependent communities may disappear, along with their associated ecosystem services. Nonetheless, rewilding can give an opportunity to tackle the issue of farmland abandonment. This chapter first depicts the historical European landscapes and the role of two natural disturbances, herbivory and fire. The importance of disturbance-dependent habitats is then highlighted

L. M. Navarro (✉) · H. M. Pereira

German Centre for Integrative Biodiversity Research (iDiv) Halle-Jena-Leipzig, Deutscher Platz 5e, 04103 Leipzig, Germany
e-mail: laetitia.navarro@idiv.de

Institute of Biology, Martin Luther University Halle-Wittenberg, Am Kirchtor 1, 06108 Halle (Saale), Germany

Centro de Biologia Ambiental, Faculdade de Ciências da Universidade de Lisboa, Campo Grande, 1749-016 Lisboa, Portugal

V. Proença
IN+, Center for Innovation, Technology and Policy Research, Environment and Energy Scientific Area, DEM, Instituto Superior Técnico, University of Lisbon, 1049-001 Lisboa, Portugal
e-mail: vania.proenca@ist.utl.pt

J. O. Kaplan
Institute of Earth Surface Dynamics, University of Lausanne, Geopolis Quartier Mouline, 1015 Lausanne, Switzerland
e-mail: jed.kaplan@unil.ch

H. M. Pereira, L. M. Navarro (eds.), *Rewilding European Landscapes,*
DOI 10.1007/978-3-319-12039-3_8, © The Author(s) 2015

by drawing attention to the alpha and beta diversity that they sustain. Finally, the chapter investigates options for rewilding abandoned land to maintain disturbance-dependent and self-sustained habitats for which we suggest active restoration in the early stages of abandonment. This may be achieved via prescribed burning and support or introduction, when necessary, of populations of wild mammals.

Keywords Disturbances · Fire regime · Disturbance-dependent habitats · Herbivory · Reintroduction · Prescribed burning

8.1 Introduction

Disturbance can be defined as "a discrete event that disrupts the structure of an ecosystem's community or population, and changes resources availability or the physical environment" (Turner 1998). Natural disturbances (i.e., not deriving from human-induced processes) are an essential process of ecosystem dynamics. Among other roles, disturbances contribute to the maintenance of ecosystem structure and nutrient cycling (Attiwill 1994; Turner 1998). More important than considering the impact of a disturbance event *per se* is to consider the regime underlying disturbances. The disturbance regime determines the landscape (Turner 1998), and is characterized by the disturbance frequency and return interval, spatial extent, intensity (energy flow per area per time) and severity (magnitude of impact).

For millennia, humans have modified ecosystems with varying intensity and over various spatial extents. These anthropogenic changes imply a modification in both the natural communities and the natural processes that cause disturbance. In particular, human activities often cause the disruption of natural regimes, either directly (e.g., livestock grazing, fire suppression) or indirectly (e.g., landscape fragmentation, introduction of exotic invasives or pests), or introduce new types of disturbance, such as pollution. Human activities can also mimic natural disturbance regimes and affect biotic communities in a similar way (Attiwill 1994). For example, the maintenance of traditional landscapes and the species-rich communities associated with them is implicitly linked with continuous ecosystem disturbance imposed by human activities.

If the regime of anthropogenic disturbances is altered, by a reduction or complete withdrawal of human activities, there is a concern that disturbance-dependent habitats and the associated communities may not be maintained. In particular, the maintenance of extensive farming systems in Europe is currently at stake due to farmland abandonment, which raises concerns about the potential effects of land-use changes on biodiversity (Rey Benayas et al. 2007). The trajectory of ecological succession after abandonment depends on several factors, but the probable shift from a moderate disturbance regime (i.e. traditional landscape mosaic) to a low or high disturbance regime is associated with the risk of habitat homogenization and decline of species richness. Thus, one of the challenges of rewilding abandoned farmland is to contribute to the maintenance of disturbance-dependent habitats.

Passive regeneration following farmland abandonment can be a long and complex process, specific to each area (see Chap. 1). It depends on the cultivation history, the time since abandonment, the availability of a "natural" seed bank, the proximity of sources of populations of species, and the requirements for natural disturbances, which will all take part in the self-sustained functioning of the restored ecosystem. When active restoration is needed, the choice of the baseline is also important (Corlett 2012), and in this regard, open land maintained quasi exclusively by (traditional) agricultural practices is a rather recent norm.

In this chapter, we first depict the European landscapes through time, from pre-modern human settlement to the progressive advent of agriculture and finally to the recent trends of agricultural abandonment. We then present two major disturbances: i.e. herbivory and fire, from both a natural history perspective and a restoration approach. We also look into the consequences of those disturbances on alpha and beta diversity levels in the landscape.

8.2 A Picture of Historical European Landscapes

An Ongoing Debate…

Describing the species, habitats, and interactions that would be present without the influence of modern humans, i.e. the pre-historical baseline, is an important step to understand natural dynamics and disturbances, and guide the restoration of self-sustaining systems (Svenning 2002; Gillson and Willis 2004; Willis and Birks 2006). However, the composition of the "pre-Neolithic landscape" (Hodder et al. 2009) is still the subject of active debate.

The Middle and Late Pleistocene interglacial can be used as proxies to describe the European pre-Neolithic landscapes, due to their similar climatic conditions and low human activity (Svenning 2002). Two contrasting pictures of lowland temperate European landscapes for these periods are described: (i) the "high-forest" hypothesis, where most of Europe was covered by forest and where the forest dynamics and the resulting availability of open-land influenced herbivore populations; (ii) the "wood-pasture" hypothesis, depicting the European landscapes as a mosaic of forest and open-land where herbivory was the main driver of openness (Vera 2000; Bradshaw et al. 2003; Birks 2005; Mitchell 2005).

Pollen records have been used to test both hypotheses and assess the degree of openness, or lack thereof, of European landscapes. Typically, the ratio between the percentage of tree pollen and non-arboreal pollen gives an indication of the openness of a landscape (Svenning 2002). Pollen records show that shade-intolerant species were present in areas both with and without evidence of large herbivores, which is in favor of the "high forest" hypothesis, in which grazers are not essential to maintaining those species (Mitchell 2005). Nonetheless, pollen and dung beetle fossil record support the idea that megaherbivores were the main keepers of openness, at least of the floodplains in Northwestern Europe (Svenning 2002), as a diverse community of dung-dependent beetles can be linked with the occurrences of large populations of herbivores (Sandom et al. 2014).

Yet, three other types of natural processes can also explain the occurrence of open areas: forest fires, windthrows and edaphic-topographic conditions (Svenning 2002; Fyfe 2007; Molinari et al. 2013). The most likely explanation is that the distribution of habitats was originally based on physical factors (Bradshaw et al. 2003), and was then enhanced and/or maintained by large herbivores, whose browsing and grazing impact delayed secondary successions.

Temporal Evolution of the European Landscape

The first hominids reached Europe from Africa in the Early Pleistocene, some 1.2–1.1 million years ago (Carbonell et al. 2008), while modern humans colonized the continent between 46,000 BP and 41,000 BP (Mellars 2006). The appropriation of new land coincided with changes in the European landscape. Nomadic hunter-gatherers started to actively manage their ecosystem with the use of fire during the Pleistocene: what started as a domestic tool (e.g. for cooking, heating, and for protection from predators) also became useful to draw game to hunting grounds, to clear travel routes, and to open space for grazers (Daniau et al. 2010; Kaplan et al. 2010; Pfeiffer et al. 2013).

The development of agriculture was the next step in humans' appropriation and management of their environment (Pereira et al. 2012). The spread of agriculture from the northern Levant and northern Mesopotamian area towards Europe has been calculated to have started between 11,550 and 9000 BP and expanded at a rate of 0.6 to 1.3 km/year, with agriculture reaching north-western Europe in 3000 years (Pinhasi et al. 2005; Ruddiman 2013). Such spread of agriculture led to a fivefold increase in the human population (Gignoux et al. 2011), which had considerable consequences on the landscape. Several models have been designed to investigate the historical evolution of this human impact. Models that do not assume a direct linear link between human density and deforestation, but also consider other factors, such as technological change, show that the rate of land appropriation was much higher in the distant than in the recent past (Ruddiman 2013; Kaplan et al. 2010). First of all, as time passed and deforestation occurred, less and less forest was left to clear. Most of all, technological improvements allowed people to produce the same amount of food on less land, which contradicts a direct link between population density and deforestation (Ruddiman 2013). Following these non-linear concepts, Kaplan et al. (2010) presented model scenarios of Holocene anthropogenic land cover change. At 8000 BP, only Mesopotamia and Turkey were showing signs of human use of the land, but by the beginning of the Iron Age at 3000 BP, up to 40% of European land could have been cleared for extensive agriculture and pastures (Fig. 8.1). Between 8000 and 3000 BP, Kaplan et al (2010) suggest that land use in Western Europe ranged from 5.5 to 6.5 ha per capita and was relatively stable.

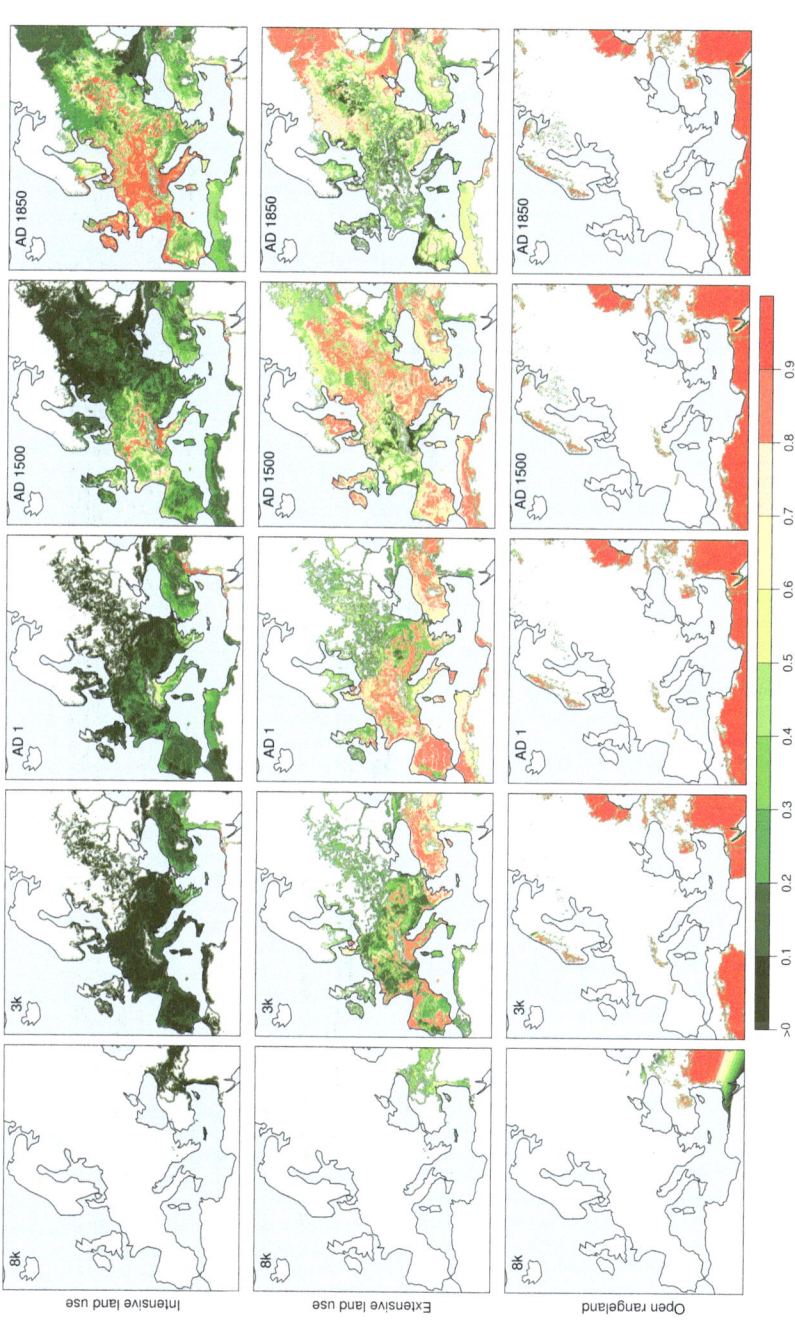

Fig. 8.1 Anthropogenic land cover change in Europe over the Holocene. Intensive land use includes land completely deforested and used for agriculture and pasture, while extensive land use includes forest-pasture, coppice and other managed forestlands. Open rangeland occurs on land that is either too cold or too dry for non-irrigated agricultural land uses. Land use is driven by estimates of past population at the country-level. (Based on Kaplan et al. 2009, 2010; Kaplan 2012)

By 2500 BP, increasing populations in most of Western Europe triggered intensification of land use (Fig. 8.1) and decrease in per capita values. Later, decreases in population resulted in major land abandonment episodes, during the Migration Period following the fall of the Western Roman Empire, and after the Black Death epidemic of AD 1350. By AD 1850, the latest preindustrial time, most of the European landscapes usable for intensive crop or pasture were deforested, and land use had dropped to values close to 0.5 ha per capita.

After the Industrial Revolution, the relationship between population and land use had become largely uncoupled. Beginning in the late eighteenth century, these "forest transitions" (e.g. Mather et al. 1998) led to abandonment of unproductive agricultural and pasture land in most European countries. Since the early 1960s, the rural population decreased by 17 % in Europe (FAOSTAT 2010), with repercussions for agricultural land-use, and both are projected to continue decreasing in the decades to come. By 2030, up to 15 % of the land cultivated in 2000 could be abandoned (e.g. Verburg and Overmars 2009; Eickhout et al. 2007) which represents 10–29 million ha of land. The areas facing the greatest likelihood of rural abandonment are remote and/or mountain areas, classified as "least favored", with marginal value for agriculture (e.g. MacDonald et al. 2000). With the withdrawal of human activities, those abandoned areas are often left without the artificial disturbances that had replaced the natural ones, centuries or millennia ago.

8.3 The Role of Natural Disturbances

Investigating the history of natural disturbances can inform researchers and managers on guidelines for restoration (Donlan et al. 2006). We identified two types of disturbances as fundamental in the maintenance of European landscapes, prior to human appropriation of the land: large herbivores and natural fire dynamics.

The Pre-Neolithic Ecosystem Engineers

Ecosystem engineers are organisms that create and/or maintain habitats, either directly or indirectly (Jones et al. 1994; Wright and Jones 2006) and thus create niches for other species. The fact of grazing and browsing is not enough to be qualified as engineering (Wright and Jones 2006). Nonetheless, the consequences of herbivory, trampling, and fertilizing, especially by large herds of megafauna, have a direct impact on the distribution of habitats (Vera 2000; Birks 2005). Small mammals are also known to have an important impacts on the vegetation, for example by disturbing the soil and modifying its physical and chemical properties (Jones et al. 1994), but this goes beyond the scope of this chapter.

During the interglacial cycles of the late Quaternary, and prior to massive extinctions, the landscapes of Europe were characterized by a rich megafauna (Bradshaw et al. 2003). The available fossil evidence can attest the presence of species in a given region, while using the impact of similar extant species as a proxy can inform on the role of extinct megaherbivores on the landscape (Corlett 2012). Nonetheless, in contrast with pollen, there are too little fossil records of pre-Neolithic large herbivores to allow for an estimate of their past densities (Bradshaw et al. 2003; Mitchell 2005), and we still lack precise knowledge regarding their behavior (Hodder et al. 2009).

The late Pleistocene megafauna of Europe (Table 8.1) resembles the one currently found in savannas, with herbivores such as proboscidae and rhinocerotidae, and large carnivores such as hyaenidae and felidae (Blondel and Aronson 1999; Vera 2000; Bradshaw et al. 2003). Globally, the group of large herbivores suffered more prehistoric extinctions than other taxa (Johnson 2009). Cyclic climatic change had typically been responsible for a regular faunal turnover, and was later combined with increased human pressure (Corlett 2012; Morrison et al. 2007), leading to several of these megafauna becoming regionally (e.g. hippopotamus), globally (e.g. woolly mammoth), and often functionally, extinct (Blondel and Aronson 1999; Bradshaw et al. 2003). Some species also suffered large range contractions, such as the elk (Morrison et al. 2007). Additionally, humans domesticated animals in the Fertile Crescent, about 10,000 years ago (Zeder 2008; Pereira et al. 2012), and as herders migrated west, increasing the area of pasture in Europe, wild herbivores were replaced by domesticated species. Since AD 1, most of the open rangeland in Europe has been under human land-use (Fig. 8.1).

Extinct and extant large herbivores can be classified according to their feeding behavior (Vera 2000; Svenning 2002; Bullock 2009): browsers (e.g. elk, straight-tusked elephants) are typically associated with tree rich areas; grazers (e.g. hippopotamus, aurochs) are in contrast associated with the occurrence of grass-rich habitat; finally, mixed feeders (e.g. red deer, wild goats) alternate between browsing and grazing (Table 8.1). The European bison, a mixed feeder, has for example been associated historically with both closed forest and semi-open habitats (Kuemmerle et al. 2012). The social structure of the herbivores (i.e. solitary, groups or herds) also provides information on the grazing and browsing pressure on the landscape (Table 8.1).

As a result, one of the most direct impacts of large herbivores on the landscape is the limitation and variation in the spatial distribution of secondary successions (Laskurain et al. 2013; Kuiters and Slim 2003). Yet, the role of herbivores goes beyond the direct impacts of browsing and grazing. For example, elephants are known to create large physical disturbances via the trampling of trees and shrubs (Jones et al. 1994), which changes their habitat, the fuel load and the local fire regime, and in return benefits light demanding plant species. The disturbance induced by the rooting behavior of wild boars favors natural forest regeneration, while being considered as damaging to grasslands (Schley et al. 2008; Sandom et al. 2013a). Large herbivores also have a role as seed-dispersers via their consumption of large quantities of forage: the low oral process of the fruits contained in this forage allows the dispersal of undamaged seeds in feces (Corlett 2012; Johnson 2009). Some seeds even need to pass through a digestive track to trigger germination. Finally, herbivore dung is important for nutrient cycling and soil fertilization (Zimov 2005).

Table 8.1 List of extant and extinct species of large herbivores present during the late Pleistocene and at some point between the early Holocene and the current time, in Europe. (Blondel and Aronson 1999; Bradshaw et al. 2003; Bullock 2009; Smith et al. 2003; Svenning 2002; Vera 2000). IUCN Status: *LC* least concern, *VU* vulnerable, *CR* critically endangered. Feeding behavior: *M* mixed feeders, *G* grazers, *B* browsers. Social structure: *G* groups, *H* herds, *S* solitary

Species	Extant in the Holocene	Extinct	Extinct Europe	Extinct locally	European IUCN Status	Feeding behavior	Social structure
Alces alces (Eurasian elk)	X			X	LC	B	S
Bison bonasus (European bison)	X			X	VU	M	H
Bison priscus (Steppe bison)		X				G	
Bison schoetensacki		X				G	
Bos primigenius (Auroch)	X	X				G	H
Bubalus murrensis (Murr water buffalo)		X				M	
Capra aegagrus (Wild goat)	X				VU	M	H
Capra ibex (Alpine ibex)	X			X	LC	M	G
Capra pyrenaica[a] (Iberian ibex)	X		X[a]	X	LC/VU	M	G
Capreolus capreolus (Roe deer)	X				LC	B	S
Capreolus suessenbornensis		X				M	
Castor fiber (Beaver)	X			X	LC	B	S
Cervus elaphus (Red deer)	X			X	LC	G	G
Coelodonta antiquitatis (Woolly rhinoceros)		X				B	
Dama clactoniana		X				M	
Dama dama (Fallow deer)	X				LC	M	H
Dicerorhinus hemitoechus (Narrow-nosed rhinoceros)		X				B	
Equus ferus (Tarpan)	X	X				G	H

Table 8.1 (continued)

Species	Extant in the Holocene	Extinct	Extinct Europe	Extinct locally	European IUCN Status	Feeding behavior	Social structure
Equus germanicus (Forest horse)	X	X				G	
Equus hydruntinus (European ass)	X	X				G	
Equus przewalskii (Przewalski's horse)	X		X			G	G
Hipparion crassum		X				B	
Hippopotamus amphibius (hippopotamus)	X		X			G	G/S
Hippopotamus antiquus (European hippopotamus)		X				M	
Mammuthus primigenius (Woolly mammoth)	X	X				G	G
Megaloceros cazioti		X				M	
Megaloceros dawkinsi (Giant deer)		X				M	
Megaloceros euryceros		X				M	
Megaloceros giganteus (Irish giant elk)	X	X				M	
Ovibos moschatus (Muskox)	X			X	LC	M	H
Ovis aries orientalis (Mouflon)	X			X	VU	M	G
Palaeoloxodon antiquus (Straight-tusked elephant)		X				M	G
Pseudodama nestii		X				M	
Rangifer tarandus (Reindeer)	X			X	LC	M	H

Table 8.1 (continued)

Species	Extant in the Holocene	Extinct	Extinct Europe	Extinct locally	European IUCN Status	Feeding behavior	Social structure
Rupicapra pyrenaica (Pyrenean chamois)	X			X	LC	M	G
Rupicapra rupicapra (Chamois)	X			X	LC	M	H
Saiga tatarica (Saiga)	X		X		CR	M	H
Soergelia elisabethae		X				M	
Stephanorhinus kirchbergensis (Merck's rhinoceros)		X				B	
Sus scrofa (Wild boar)	X			X	LC	M	G
Ursus spelaeus (Cave bear)[b]		X				M	S

[a] There are four subspecies of *Capra pyrenaica*, two of which are extinct—*C. pyrenaica lusitanica* and *C. pyrenaica pyrenaica*—while two are extant—*C. pyrenaica victoriae* and *C. pyrenaica hispanica*
[b] Cave bears had a predominantly herbivorous diet

Fire Dynamics

Fire is a critical component in the functioning of many ecosystems. It maintains and shapes vegetation structure and biotic communities, promotes natural regeneration and habitat diversity, takes part in biogeochemical cycles, and can influence soil properties and water functions (Thonicke et al. 2001; Bond and Keeley 2005). Unlike grazing, fires consume both dead and living material and do not discriminate between edible and non-edible plants (Bond and Keeley 2005), but may act as a selective pressure over fire resistant traits (Pausas and Bradstock 2007; Pausas et al. 2006).

Fire-dependent systems cover about 53% of the world's terrestrial surface (Shlisky et al. 2007). These systems evolved in the presence of fire and depend on this disturbance to maintain their structure and composition (e.g., Mediterranean forests and boreal forests), with fire regimes characterized by their frequency, intensity, seasonality, and specific to each ecosystem. In addition, 22% of the world's terrestrial area is covered by fire-sensitive ecosystems, where fire plays a minor role in maintaining ecosystem structure and composition (e.g., broadleaved and mixed forests in the Alps), 15% is covered by fire-independent ecosystems, where fire is not an evolutionary force due to the scarcity of fuel or ignition sources (e.g. tundra), and the remaining 10% are not yet classified (Shlisky et al. 2007).

In Europe, natural fire regimes are mainly of two types: (i) intense and large, and (ii) cool and small (Archibald et al. 2013). The former type is typical of Mediterranean and boreal ecosystems, where large crown fires of high intensity return at intervals that can span from a decade, in particular in Mediterranean regions, to more than a century (Archibald et al. 2013). The latter type occurs interspersed with the first type, in the same biomes, and is associated with surface fires burning litter fuels (Archibald et al. 2013). However, due to a long history of human presence, many ecosystems in Europe, including fire-sensitive systems, present altered fire-regimes resulting from land-use changes and anthropogenic fire management (Shlisky et al. 2007; Archibald et al. 2013; Molinari et al. 2013). Current yearly fire occurrence in Europe ranges from less than five per NUTS3[1] to nearly a hundred in areas of the Mediterranean region, which also presents the largest average of area burned yearly, with over 10,000 ha/year in some NUTS (European Commission 2010). Four types of areas can be identified in Europe, based on their fire regimes, when combining both the occurrences of fire and the average area burned in each NUTS3 (Fig. 8.2). Central France, North-Eastern Germany, and most of Romania present small fire regimes, with few fires (<20 per year) and little area burned (<35 ha). Poland, most of the Baltic and Scandinavian countries are areas with relatively high occurrences of fire (>50 per year) but small area burned (<35 ha). In contrast, most of Bulgaria and Greece are regions where a small number of fires (<20 per year) are sufficient to burn large areas (>115 ha). Finally, Southern Italy, Croatia and the Iberian Peninsula are areas with both high fire frequency (>50 per year) and large areas burned (>115 ha).

---

[1] Third level of the EU Nomenclature of Territorial Units for Statistics.

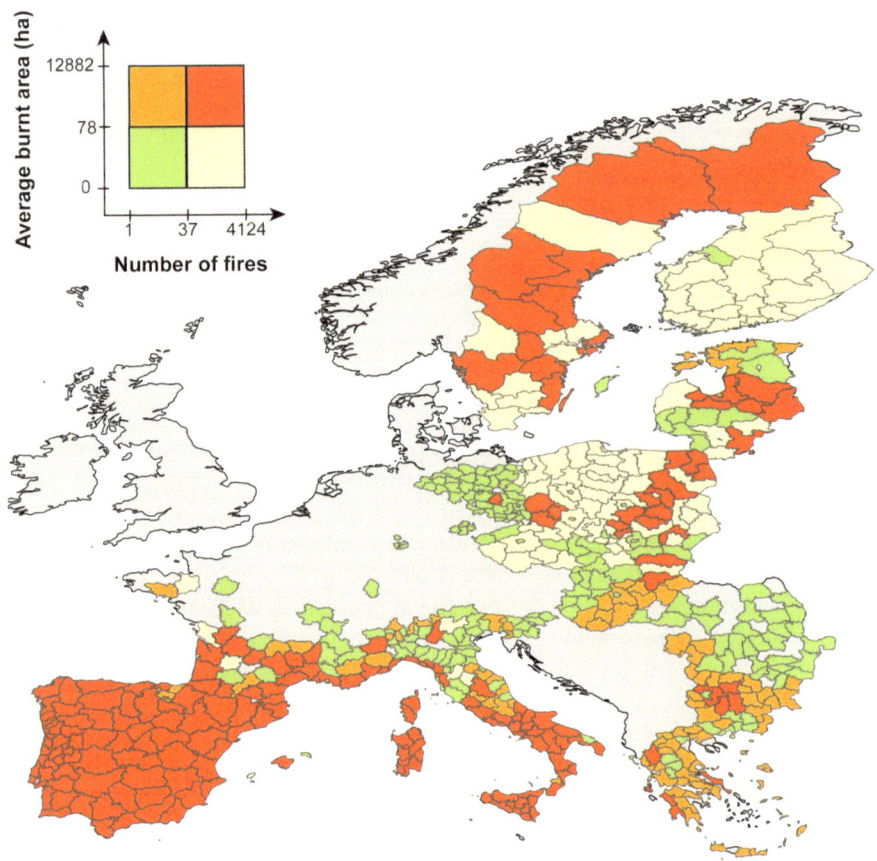

Fig. 8.2 Occurrence and intensity of fires in Europe over the 2005–2010 period. The average yearly occurrence of fire and average area burned (ha) for the 2005–2010 period, per NUTS3 administrative unit were calculated, only including NUTS3 for which data were available for at least 4 years. For both metrics, the data were split in two groups around the median value. The double color ramp allows to identify areas with high number of fire but low area burned (*yellow*), areas with low occurrence of fire but large burned areas (*orange*), areas with few fire and small areas burned (*green*), and areas with both high occurrence of fire and large burned areas (*red*). (Source: EFFIS for the fire data (European Commission 2010) and © EuroGeographics for the map of administrative boundaries)

Fire suppression is a common land management policy implemented to protect human communities and land (Shlisky et al. 2007; Fernandes 2013) but it also promotes fuel accumulation in fire-dependent systems and increases the risk of large and intense fires (Proença et al. 2010; Fernandes 2013). On the other hand, fire has also been extensively used as a tool to clear landscapes and reduce fire risk. In Europe, anthropogenic fires are often more frequent than natural fires. High frequency fire regimes can cause species community impoverishment, through the exclusion of fire sensitive species and the promotion of fire resilient species that can endure frequent fires, and it can also cause extensive soil degradation and nutrient loss (Thonicke et al. 2001). This is particularly true for Mediterranean ecosystems, where 93 % of fire regimes are considered to be in a degraded or very degraded state (Shlisky et al. 2007).

Today, farmland abandonment is driving further changes in fire regimes across Europe, particularly in Southern Europe, with potential impacts for biodiversity and ecosystem services (Mouillot et al. 2005; Bassi et al. 2008; Proença and Pereira 2010). Where the number of ignitions is not a limiting factor, which is true in many regions under farmland abandonment (Bassi et al. 2008; Ganteaume et al. 2013), climate and fuel availability will be the main determinants of future changes to the fire regime. In high-productivity ecosystems with a high level of humidity, such as temperate broadleaved forests, fires will be limited by climate and humidity level, and less responsive to changes in fuel accumulation, since fuel is already a non-limiting factor (Pausas and Ribeiro 2013). Vegetation will be more susceptible to fire during warmer seasons following droughts, when the existing fuel is more flammable (Proença et al. 2010; Pausas and Ribeiro 2013). In low-productivity ecosystems, such as arid Mediterranean scrublands, fuel is the main limiting factor and will be the main driver of shifts in the fire regime (Pausas and Fernández-Muñoz 2012; Pausas and Ribeiro 2013). Recent trends in the Western Mediterranean Basin support the above predictions (Pausas and Fernández-Muñoz 2012). In this region, fields used to be grazed, frequently burned (small scale) and cleared for farming and timber (Proença and Pereira 2010), limiting fuel availability. The rural exodus since the mid-twentieth century led to shrub encroachment and afforestation with fire-prone species, and resulted in more frequent, more intense and larger fires. Today, increased fuel load and spatial continuity are driving a shift in the fire regime, which is becoming more responsive to drought, similar to high-productivity ecosystems (Pausas and Fernández-Muñoz 2012). In the future, the response of fire regime to changes in climatic variables, such as precipitation, is expected to be non-linear (Batllori et al. 2013): while a small decrease in annual precipitation may increase probability of fire, a large decrease may lead to the inverse response due to a drop in ecosystem productivity, leading the system back to a fuel-limited fire regime.

8.4 Disturbances and Diversity

Traditional landscapes in Europe, in particular High Nature Value (HNV) farm-land areas, are acknowledged for their high species richness and conservation value (Blondel and Aronson 1999; MacDonald et al. 2000; EEA 2004). Species diversity patterns in traditional landscapes are likely to be different from what would be found in non-modified (primary) landscapes (Blondel and Aronson 1999). When the total number of species is considered, a higher richness of species at the habitat patch scale (i.e., α-diversity) is expected in traditional landscapes due to species being able to use more than one habitat and due to the high density of habitat edges, which facilitates inter-patch movements and therefore leads to a higher species turnover in space and time (Proença and Pereira 2013; Guilherme and Pereira 2013). Note that even with inter-patch movements, each habitat type will support a distinct community of species due to differences in species abundances and due to the existence of strict habitat specialists. As a result, the α-diversity is probably lower in the case of specialist species in traditional mosaics due to the effect of habitat fragmentation and their low tolerance to the conditions found in other habitats (Proença and Pereira 2013). For instance, the diversity of forest species is lower in fragmented forest patches than in an area of similar size in continuous habitat (Proença 2009). Regarding species turnover (i.e., β-diversity), traditional landscapes can have a higher turnover than former undisturbed land (Blondel and Aronson 1999), due to their mosaic structure. However, the soundness of this assumption depends on the scale of the analysis (see Chap. 6). For example, one can predict that the replication of the traditional habitat mosaic across large spatial scales results in a higher similarity of (modified) habitats, which promotes the presence of similar communities across large areas. Finally, the effect of these changes on the total number of species found in the landscape (i.e., γ-diversity) is less straightforward. Indeed, whilst several species suffered declines or even extinctions due to habitat destruction or modification (e.g., bear, auroch), other species benefited from these changes and proliferated in the human-modified habitats (e.g., farmland birds). Moreover, starting in the earliest Neolithic, farmers continually and intentionally introduced new species to European ecosystems (Blondel and Aronson 1999). They also did so unintentionally as a result of species dispersal by animal herds along transhumance routes (e.g. Poschlod et al. 1998). Both of these activities thus increased the regional species pool, though globally richness declined due to extinctions.

Diversity and Intermediate Disturbance

The intermediate disturbance hypothesis (Connell 1978) and the diversity-disturbance hypothesis (Huston 1979) are often used to explain the ecological mechanisms determining the high diversity of species found in traditional landscapes (e.g. Blondel 2006): species diversity peaks when communities are exposed to moderate disturbance, in terms of frequency, extent and intensity. This occurs because moderate disturbance (e.g., moderate grazing) creates discontinuities in the ecosystem that allow the maintenance of early successional species while preventing dominance of more competitive species, hence keeping the ecosystem in a transitional state between early and steady-state communities. The management of traditional landscape mosaics (Fig. 8.3), with low intensity farming, moderate grazing and maintenance of forest patches, is often described as an example of intermediate disturbance, and therefore as a promoter of species diversity (Ostermann 1998; Henle et al. 2008). Nonetheless, peaked relationships between species richness and disturbance are not the rule across ecology studies (Mackey and Currie 2001). Peaked curves are more commonly reported by studies covering small spatial scales and in the presence of natural disturbance regimes (Mackey and Currie 2001). In

Fig. 8.3 Minifundia system in a mountain landscape in Northwestern Portugal. (Photo credit: Vânia Proença)

addition, the relationship between taxa richness and the intensity of anthropogenic disturbance regimes is often non-significant (Mackey and Currie 2001), increasing the challenge of predicting the impacts of altered regimes of disturbances on biodiversity.

Effects of Land-Use Change on Disturbance Regimes

Land-use changes caused by rural abandonment can create the conditions for an increase in the frequency and intensity of disturbance events, in particular higher fire risk due to fuel accumulation and shrub encroachment, but may also result in fewer disturbances if disturbance agents, such as domestic grazers or browsers, become residual or even disappear. The trajectory of secondary succession after abandonment depends on several interacting factors and ecological filters, such as the pool of colonizer species in the surrounding landscape, their ability to colonize abandoned patches, soil quality, and, of course, disturbance regime (Cramer 2007). Disturbances will not only exert a selective pressure on community assembly, but will also respond to community structure and composition.

In landscapes where tree density is very low, such as some Mediterranean landscapes, there is a high probability of shrub encroachment after farmland abandonment due to seed limitation, predatory pressure over oak acorns and deficient abiotic conditions, such as poor soils (Acácio et al. 2007). Wildfire will further promote shrub dominance, due to many shrubs' resprouting ability. Wildfires may hence establish a reinforcing feedback loop, leading to community homogenization and a decline in diversity at all scales (Proença and Pereira 2010).

A different trajectory can be anticipated in landscapes with a higher tree density, such as semi-natural grasslands in northern Europe (Eriksson et al. 2002). There, seed availability and dispersal are not limiting factors and forest is able to colonize and regenerate in relatively short time. With an expected low disturbance regime, forest can expand, which would decline habitat heterogeneity. Some species, such as grassland specialists, will show strong reductions in abundance or even go locally extinct. Impacts at the landscape level will depend on species ability to persist in alternative habitats such as forest edges or heathlands (Proença and Pereira 2013).

The above examples describe abandoned patches in a fairly homogenous landscape matrix with either a low or high tree density. In a heterogeneous landscape with a more balanced cover of different habitats and a variety of edaphic-topographic conditions, scenarios would be different given the diversity of local responses to changes in disturbance regime. Habitat diversity will not only counteract landscape homogenization, but also provide alternative habitats for species affected by farmland abandonment, thus reducing the impact of land-use change on species diversity. The persistence of those species in the landscape will then depend on the maintenance of those alternative habitats, either by natural processes, such as herbivory by wild ungulates, or through assisted processes, such as prescribed fire or herbivore re-introduction. Such restoration approaches, either passive or assisted, are an important open question in rewilding research.

8.5   Maintaining Disturbance-Dependent Habitats

Wild Herbivores: Natural (Re)colonization or (Re)introduction?

Today, only 16 % of the Palearctic region, including Europe, contains areas occupied by relatively undisturbed large mammal faunas, i.e., species that have not undergone major changes in range between AD 1500 and the present (Morrison et al. 2007). This figure does not even consider the number of species that went extinct early in the Holocene (Table 8.1). There is also a clear regional difference when looking at

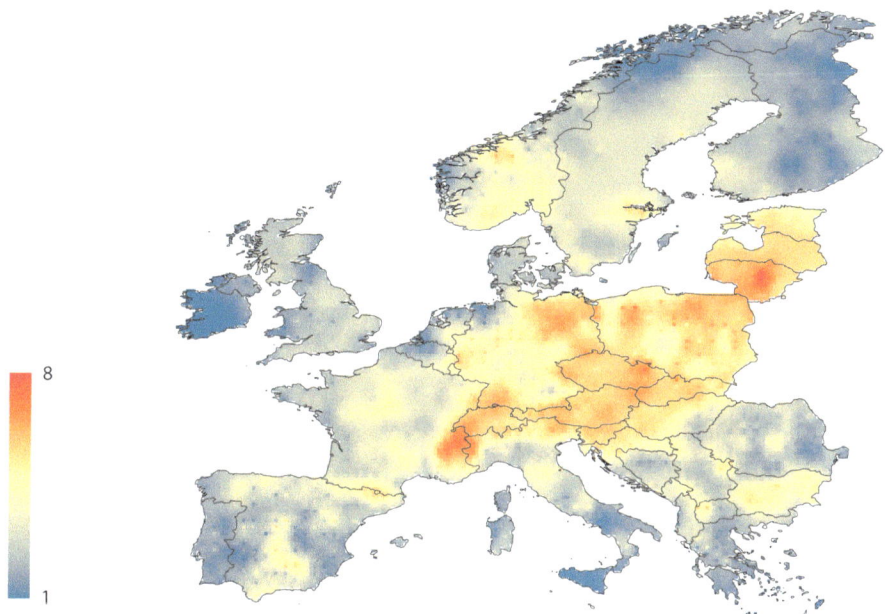

Fig. 8.4 Species richness for extant large herbivores of Europe—See Table 8.1 for the list of species. Map obtained using Inverse Distance Weighting (weight=2) on the atlas data. (Source: Atlas of European Mammals, Mitchell-Jones et al. 1999)

the current species richness of large herbivores in Europe (Fig. 8.4): countries of central Europe present the highest diversity, while the Westernmost countries have low richness. Species rich areas, with lower human densities and less pressure on the land, could become "sources" for natural re-colonization. This has already been documented for some species of large herbivores that show substantial increases in their populations since the 1960s (Table 8.2). Though legislation and conservation measures largely contributed to it (Deinet et al. 2013), rural depopulation and the associated reduced human pressure, both direct (e.g. less hunting) and indirect (e.g. more land available), can also explain the phenomena (Table 8.2). Wild populations can also benefit from the absence of competitor and predator species (Bradshaw et al. 2003), though unregulated population growth could become an issue, e.g. if their pressure on the land is too high.

In cases where the local richness of wild herbivores is low, as for example in Western European countries (Fig. 8.4), species can be introduced to restore ecosystem functioning (Sandom et al. 2013b). That is, provided that their functional role is left unattended (Lipsey and Child 2007), and that the abandoned land meets their requirement in natural resources. A study on fenced populations of wild boar showed that their rooting behavior can create germination niches (Sandom et al. 2013a) and contribute to forest regeneration. However, they can also be detrimental to the established trees when bark stripping and uprooting (Sandom et al. 2013b). Reintroducing ecosystem engineers to restore and/or maintain disturbance dependent habitats can also be more time and cost effective than man-made restoration (Byers et al. 2006; Sandom et al. 2013a). Moreover, provided that the re-introduced species present charismatic values, their presence could facilitate the acceptation of a rewilding project by the public (Lipsey and Child 2007; Kuemmerle et al. 2012). The reintroduction of wild grazers can also be assessed positively from the standpoint of ecosystem services, based on the existence value of the megafauna (Proença et al. 2008) and associated cultural services (e.g. tourism, hunting, and see Chap. 3).

Nonetheless, a balance must be maintained when considering the (re)introduction of herbivores and many potential challenges should be raised and discussed (Seddon et al. 2007; Corlett 2012; IUCN 2013). First, which species should be reintroduced? When taxon substitutions are needed for ecological replacement (IUCN 2013), researchers' opinions are divided, ranging from releasing breeds of domesticated animals, to the reintroduction of extant relatives of long gone species (e.g. Donlan et al. 2006). Releasing animals also raises the question of increasing the risk of conflicts between local human populations and "wildlife" (e.g. Enserink and Vogel 2006; Goulding and Roper 2002), which could be more easily accepted if the species was progressively, and naturally, recolonizing an area. For reintroduced domestic species (e.g. horses), a legal framework on the liability of the organization that performed the reintroduction is also missing, for instance in cases of damages or accidents. Finally, an overabundance of certain species can have detrimental effects, especially when the natural predator guild is absent and cannot regulate the populations (see Chap. 4), yet no specific guidelines are designed for the natural control of reintroduced populations (IUCN 2013). For instance, the large populations of browsers in the Scottish Highlands, where large carnivores have been extinct for centuries, currently limit the natural forest regeneration (Sandom et al. 2013b).

Prescribed Burning

Fire can be used as a tool in landscape management for two main intents: to control fire risk and the intensity of wildfires, and to manage landscape structure and biodiversity. Prescribed fires are often used as a preventive measure to control fuel load and fire intensity (e.g. Fernandes 2013). In addition, the combination of different fire regimes can be used to maintain landscape heterogeneity and habitat for species dependent on different ecosystem successional stages (Driscoll et al. 2010). In regions where fire risk and shrub encroachment are paired threats to biodiversity conservation, fire can be used as a tool to approach both problems (Moreira and Russo 2007).

Nonetheless, the use of prescribed fires can also raise some conservation issues. For instance, prescribed fires are performed during the wet season (winter to spring) when there is a low risk of fire spreading, while natural fires occur during dry days, especially in summer. This divergence in fire season can negatively impact species that reproduce in spring (van Andel and Aronson 2012), such as ground nesting birds, but also the persistence of plant species, for example by causing premature

Table 8.2 Population trends for large herbivores in Europe and main reasons for recovery of the populations. (Based on Deinet et al. 2013)

Species	Population estimates (year)[a]	Population increase (1960–2005) (%)	Ranked causes for increase[b]				Natural recolonization observed?
			Legislation	Conservation measures (including reintroductions)	Reduced human pressure (e.g. land abandonment, decreased hunting)	Change in environmental conditions	
Alces alces (Elk)	719,810 (2004/2005)	210	3	1	2	4	X
Bison bonasus (European bison)	2759 (2011)	+3000		1		2	
Capra ibex (Alpine ibex)	36,780 (2004/2005)	475	1	2		3	
Capra pyrenaica (Iberian ibex)	>50,000 (2002)	875	2	1	3		
Capreolus capreolus (Roe deer)	9,860,049 (2005)	240	3	2	1	4	X
Castor fiber (Beaver)	>337,539 (2003–2012)	>14,000	1	2	3		X
Cervus elaphus (Red deer)	2,443,035 (2002–2010)	400	2	1	3	4	X
Rupicapra pyrenaica (Pyrenean chamois)	69,100 (2008)	550	1	2	3		
Rupicapra rupicapra (Chamois)	485,580 (2004/2005)	85	4	1	2	3	
Sus scrofa (Wild boar)	3,994,133 (2004–2012)	400		1	3	2	

[a] Some population estimates are obtained by summing values of national assessments performed in different years, hence a time interval instead of a year of assessment in some cases

[b] Ranking based on Deinet et al. (2013), with "1" being the most relevant and "4" the least relevant. The number of observed causes of increase varies from species to species

seed release, or by destroying seedlings of annual plants before they create a seed bank (Whelan 1995; Bowman et al. 2013). Another issue is the implications of prescribed fires for climate change mitigation. Large scale prescribed fires may aggravate climate change, due to the emission of greenhouse gases and aerosol particles (Russell-Smith et al. 2009; Fernandes et al. 2013). While more research is needed to understand the effects of prescribed burning on carbon cycle (Fernandes 2013), it is also accepted that well planned prescribed burning prevents larger losses of carbon to the atmosphere by reducing wildfire risk (Bowman et al. 2013; Fernandes 2013). Finally, defining the regime of prescribed fires can be challenging (Whelan 1995; van Andel and Aronson 2012). Replicating natural fire regimes may not be possible, due to the lack of historic information. It may even not be advisable, given changes in landscape structure and, in some areas, in local climate, which may lead to unpredicted responses to fire (Driscoll et al. 2010). Therefore induced fire regimes should be planned to meet the desired outcomes instead of trying to mimic the parameters of natural fire regimes (Whelan 1995). In particular, in a rewilding context, fire dynamics should only be managed, or "assisted" in the early stages post-abandonment in order to facilitate the restoration of natural fire regimes.

8.6   Concluding Remarks

Millennia of human activities have progressively replaced natural disturbances, such as herbivory and fire, to shape the European landscapes. Maintaining disturbance-dependent habitats after the withdrawal of those human activities is a difficult restoration process. It can be guided by knowledge of the past (Vera 2000; Gillson and Willis 2004; Willis and Birks 2006; Sandom et al. 2013b), and by improving our ability to understand ecosystem dynamics and projecting potential restoration pathways. This means identifying the most desirable outcome in terms of both biodiversity and resilience. Nonetheless, besides human impacts on the landscapes, other biotic and abiotic alterations have also led to the current ecosystem composition. The climate has changed during the past millennium and some species have gone extinct while others have invaded, all these changes influencing ecological processes (Gillson and Willis 2004; Hodder et al. 2009). The interaction between human pressure and natural changes (e.g. non-anthropogenic climatic changes) could also have led to the crossing of tipping points (Gillson and Willis 2004; Kaplan et al. 2010; Leadley et al. 2014). Returning the landscapes to their historical conditions would thus be unachievable, if even desirable. This means that the baseline must shift, not only for the policy makers and the public who attribute cultural values to a relatively recent landscape (Vera 2009), but also for scientists and conservationist, some of which, on the contrary, having too long of a memory of the European landscape.

An additional concern emerges with farmland abandonment when herbivores also become functionally extinct, following a decrease in agricultural activities (Donlan et al. 2006), while the artificial fire regime is altered. Hence, in the early stages after land abandonment, the "restoration goals" must be defined to determine the set of biotic and abiotic factors that might be managed (Byers et al. 2006). Supporting local populations of wild herbivores, reintroducing them in places where they are absent and using prescribed burning can constitute the first steps towards restoring ecological processes.

For instance, the choice between natural recolonizations, reinforcement of local populations or reintroductions will depend on the current distribution and abundances of the herbivore communities. In areas of Central Europe, one might expect that the diversity of herbivores is high enough to allow for recolonizations, while in Western and Southern Europe, active introduction might be needed (Fig. 8.4). In all cases, conservation measures, legislation and reduced human pressure are necessary for the establishment of viable populations (Table 8.2).

When rewilding is meant to lead to ecological restoration, reintroductions should be one of the tools rather than a goal per se. Moreover, historical baselines should be treated as guidelines, not as objectives. In other words, rather than focusing on the conservation of a given set of species or habitats, rewilding will focus on the restoration and conservation of natural processes, with human intervention reduced to its minimum.

Acknowledgments We thank Thomas Merckx and Christopher Sandom for insightful comments on earlier versions of the manuscript. We are also thankful to the Societas Europaea Mammalogica for sharing their Atlas data. L.M.N. and V.P. were supported by fellowships from the FCT (SFRH/BD/62547/2009 and SFRH/BPD/80276/2011). J.O.K. was supported by a grant from the European Research Council (COEVOLVE, 313797).

References

Acácio, V., Holmgren, M., Jansen, P. A., & Schrotter, O. (2007). Multiple recruitment limitation causes arrested succession in Mediterranean Cork Oak Systems. *Ecosystems, 10*(7), 1220–1230.

Archibald, S., Lehmann, C. E. R., Gómez-Dans, J. L., & Bradstock, R. A. (2013). Defining pyromes and global syndromes of fire regimes. *Proceedings of the National Academy of Sciences, 110*(16), 6442–6447.

Attiwill, P. M. (1994). The disturbance of forest ecosystems: The ecological basis for conservative management. *Forest Ecology and management, 63*(2), 247–300.

Bassi, S., Kettunen, M., Kampa, E., & Cavalieri, S. (2008). *Forest fires: causes and contributing factors to forest fire events in Europe.* Study for the European Parliament Committee on Environment, Public Health and Food Safety under contract IP/A/ENVI/FWC/2006-172/LOT1/C1/SC10. Institute for European Environmental Policy, London, United Kingdom. 56p.

Batllori, E., Parisien, M.-A., Krawchuk, M. A., & Moritz, M. A. (2013). Climate change-induced shifts in fire for Mediterranean ecosystems. *Global Ecology and Biogeography, 22*(10), 1118–1129.

Birks, H. J. B. (2005). Mind the gap: How open were European primeval forests? *Trends in Ecology and Evolution, 20*(4), 151–154.

Blondel, J. (2006). The "Design" of Mediterranean landscapes: A millennial story of humans and ecological systems during the historic period. *Human Ecology, 34*(5), 713–729.

Blondel, J., & Aronson, J. (1999). *Biology and wildlife of the Mediterranean Region.* Oxford University Press, New York, USA.

Bond, W. J., & Keeley, J. E. (2005). Fire as a global "herbivore": The ecology and evolution of flammable ecosystems. *Trends in Ecology and Evolution, 20*(7), 387–394.

Bowman, D. M. J. S., O'Brien, J. A., & Goldammer, J. G. (2013). Pyrogeography and the global quest for sustainable fire management. *Annual Review of Environment and Resources, 38*(1), 57–80.

Bradshaw, R. H., Hannon, G. E., & Lister, A. M. (2003). A long-term perspective on ungulate-vegetation interactions. *Forest Ecology and Management, 181*(1–2), 267–280.

Bullock, D. J. (2009). What larger mammals did Britain have and what did they do? *British Wildlife, 20*(5), 16–20.

Byers, J. E., Cuddington, K., Jones, C. G., Talley, T. S., Hastings, A., Lambrinos, J. G., Crooks, J. A., & Wilson, W. G. (2006). Using ecosystem engineers to restore ecological systems. *Trends in Ecology and Evolution, 21*(9), 493–500.

Carbonell, E., Bermúdez de Castro, J. M., Parés, J. M., Pérez-González, A., Cuenca-Bescós, G., Ollé, A., Mosquera, M., Huguet, R., van der Made, J., Rosas, A., Sala, R., Vallverdú, J., García, N., Granger, D. E., Martinón-Torres, M., Rodríguez, X. P., Stock, G. M., Vergès, J. M., Allué, E., Burjachs, F., Cáceres, I., Canals, A., Benito, A., Díez, C., Lozano, M., Mateos, A., Navazo, M., Rodríguez, J., Rosell, J., & Arsuaga, J. L. (2008). The first hominin of Europe. *Nature, 452*(7186), 465–469.

Connell, J. H. (1978). Diversity in tropical rain forests and coral reefs. *Science, 199*(4335), 1302–1310.

Corlett, R. T. (2012). The shifted baseline: Prehistoric defaunation in the tropics and its consequences for biodiversity conservation. *Biological Conservation, 163,* 13–21.

Cramer, V. A. (2007). Old fields as complex systems: New concepts for describing the dynamics of abandoned farmland. In V. A. Cramer & R. J. Hobbs (Eds.), *Old fields: Dynamics and restoration of abandoned farmland* (p. 334). Washington, DC: Island Press.

Daniau, A. L., d' Errico, F., & Goñi, M. F. S. (2010). Testing the hypothesis of fire use for ecosystem management by Neanderthal and Upper Palaeolithic modern human populations. *PloS ONE, 5*(2), e9157.

Deinet, S., Ieronymidou, C., McRae, L., Burfield, I. J., Foppen, R. P., Collen, B., & Bohm, M. (2013). *Wildlife comeback in Europe: The recovery of selected mammal and bird species.* Final report to Rewilding Europe by ZSL. London: BirdLife International and the European Bird Census Council.

Donlan, C. J., Berger, J., Bock, C. E., Bock, J. H., Burney, D. A., Estes, J. A., Foreman, D., Martin, P. S., Roemer, G. W., & Smith, F. A. (2006). Pleistocene rewilding: An optimistic agenda for twenty-first century conservation. *American Naturalist, 168*(5), 660–681.

Driscoll, D. A., Lindenmayer, D. B., Bennett, A. F., Bode, M., Bradstock, R. A., Cary, G. J., Clarke, M. F., Dexter, N., Fensham, R., Friend, G., Gill, M., James, S., Kay, G., Keith, D. A., MacGregor, C., Russell-Smith, J., Salt, D., Watson, J. E. M., Williams, R. J., & York, A. (2010). Fire management for biodiversity conservation: Key research questions and our capacity to answer them. *Biological Conservation, 143*(9), 1928–1939.

EEA. (2004). *High nature value farmland: Characteristics, trends and policy challenges.* Copenhagen: European Environment Agency.

Eickhout, B., Van Meijl, H., Tabeau, A., & Van Rheenen, T. (2007). Economic and ecological consequences of four European land use scenarios. *Land use policy, 24*(3), 562–575.

Enserink, M., & Vogel, G. (2006). The carnivore comeback. *Science, 314*(5800), 746.

Eriksson, O., Cousins, S. A. O., & Bruun, H. H. (2002). Land-use history and fragmentation of traditionally managed grasslands in Scandinavia. *Journal of Vegetation Science, 13*(5), 743–748.

European Commission. (2010). *Forest Fires in Europe 2009.* Luxemburg: Office for Official Publications of the European Communities.

FAOSTAT. (2010). Data. http://faostat.fao.org. Accessed 1 March 2011

Fernandes, P. M. (2013). Fire-smart management of forest landscapes in the Mediterranean basin under global change. *Landscape and Urban Planning, 110,* 175–182.

Fernandes, P. M., Davies, G. M., Ascoli, D., Fernández, C., Moreira, F., Rigolot, E., Stoof, C. R., Vega, J. A., & Molina, D. (2013). Prescribed burning in southern Europe: Developing fire management in a dynamic landscape. *Frontiers in Ecology and the Environment, 11*(s1), e4–e14.

Fyfe, R. (2007). The importance of local-scale openness within regions dominated by closed woodland. *Journal of Quaternary Science, 22*(6), 571–578.

Ganteaume, A., Camia, A., Jappiot, M., San-Miguel-Ayanz, J., Long-Fournel, M., & Lampin, C. (2013). A review of the main driving factors of forest fire ignition over Europe. *Environmental Management, 51*(3), 651–662.

Gignoux, C. R., Henn, B. M., & Mountain, J. L. (2011). Rapid, global demographic expansions after the origins of agriculture. *Proceedings of the National Academy of Sciences, 108*(15), 6044–6049.

Gillson, L., & Willis, K. J. (2004). "As Earth's testimonies tell": Wilderness conservation in a changing world. *Ecology Letters, 7*(10), 990–998.

Goulding, M. J., & Roper, T. J. (2002). Press responses to the presence of free-living Wild Boar (*Sus scrofa*) in southern England. *Mammal Review, 32*(4), 272–282.

Guilherme, J. L., & Pereira, H. M. (2013). Adaptation of bird communities to farmland abandonment in a mountain landscape. *PLoS ONE, 8*(9), e73619.

Henle, K., Alard, D., Clitherow, J., Cobb, P., Firbank, L., Kull, T., McCracken, D., Moritz, R. F. A., Niemelä, J., Rebane, M., Wascher, D., Watt, A., & Young, J. (2008). Identifying and managing the conflicts between agriculture and biodiversity conservation in Europe-a review. *Agriculture, Ecosystems & Environment, 124*(1-2), 60–71.

Hodder, K. H., Buckland, P. C., Kirby, K. J., & Bullock, J. M. (2009). Can the Pre-Neolithic provide suitable models for re-wilding the landscape in Britain? *British Wildlife, 20*(5), 4–15.

Huston, M., (1979). A general hypothesis of species diversity. *American Naturalist, 113*(1), 81–101.

IUCN. (2013). *Guidelines for reintroductions and other conservation translocations. Version 1.0., Gland.* Switzerland: IUCN Species Survival Commission.

Johnson, C. N. (2009). Ecological consequences of late Quaternary extinctions of megafauna. *Proceedings of the Royal Society B: Biological Sciences, 276*, 2509–2519.

Jones, C. G., Lawton, J. H., & Shachak, M., (1994). Organisms as ecosystem engineers. *Oikos, 69*, 373–386.

Kaplan, J. O. (2012). Integrated modeling of Holocene land cover change in Europe. *Quaternary International, 279*, 235–236.

Kaplan, J. O., Krumhardt, K. M., & Zimmermann, N. (2009). The prehistoric and preindustrial deforestation of Europe. *Quaternary Science Reviews, 28*(27–28), 3016–3034.

Kaplan, J. O., Krumhardt, K. M., Ellis, E. C., Ruddiman, W. F., Lemmen, C., & Goldewijk, K. K. (2010). Holocene carbon emissions as a result of anthropogenic land cover change. *The Holocene, 21*(5), 775–791.

Kuemmerle, T., Hickler, T., Olofsson, J., Schurgers, G., & Radeloff, V. C. (2012). Refugee species: Which historic baseline should inform conservation planning? *Diversity and Distributions, 18*(12), 1258–1261.

Kuiters, A. T., & Slim, P. A. (2003). Tree colonisation of abandoned arable land after 27 years of horse-grazing: The role of bramble as a facilitator of oak wood regeneration. *Forest Ecology and Management, 181*(1–2), 239–251.

Laskurain, N. A., Aldezabal, A., Olano, J. M., Loidi, J., & Escudero, A. (2013). Intensification of domestic ungulate grazing delays secondary forest succession: Evidence from exclosure plots. *Journal of Vegetation Science, 24*(2), 320–331.

Leadley, P., Proença, V., Fernández-Manjarrés, J., Pereira, H. M., Alkemade, R., Biggs, R., Bruley, E., Cheung, W., Cooper, D., Figueiredo, J., Gilman, E., Guénette, S., Hurtt, G., Mbow, C., Oberdorff, T., Revenga, C., Scharlemann, J. P. W., Scholes, R., Smith, M. S., Sumaila, U. R., & Walpole, M. (2014). Interacting regional-scale regime shifts for biodiversity and ecosystem services. *BioScience, 64*(8), 665–679.

Lipsey, M. K., & Child, M. F. (2007). Combining the fields of reintroduction biology and restoration ecology. *Conservation Biology, 21*(6), 1387–1390.

MacDonald, D., Crabtree, J. R., Wiesinger, G., Dax, T., Stamou, N., Fleury, P., Gutierrez Lazpita, J., & Gibon, A. (2000). Agricultural abandonment in mountain areas of Europe: Environmental consequences and policy response. *Journal of Environmental Management, 59*(1), 47–69.

Mackey, R. L., & Currie, D. J. (2001). The diversity-disturbance relationship: Is it generally strong and peaked? *Ecology, 82*(12), 3479–3492.

Mather, A. S., Needle, C. L., & Fairbairn, J. (1998). The human drivers of global land cover change: The case of forests. *Hydrological processes, 12*(13–14), 1983–1994.

Mellars, P. (2006). A new radiocarbon revolution and the dispersal of modern humans in Eurasia. *Nature, 439*(7079), 931–935.

Mitchell, F. J. G. (2005). How open were European primeval forests? Hypothesis testing using palaeoecological data. *Journal of Ecology, 93*(1), 168–177.

Mitchell-Jones, A. J., Amori, G., Bogdanowicz, W., Krystufek, B., Reijnders, P. J. H., Spitzenberger, F., Stubbe, M., Thissen, J. B. M., Vohralik, V., & Zima, J. (1999). *The atlas of European mammals*. http://www.european-mammals.org/. Accessed 27th July 2013.

Molinari, C., Lehsten, V., Bradshaw, R. H. W., Power, M. J., Harmand, P., Arneth, A., Kaplan, J. O., Vannière, B., & Sykes, M. T. (2013). Exploring potential drivers of European biomass burning over the Holocene: A data-model analysis. *Global Ecology and Biogeography, 22*(12), 1248–1260.

Moreira, F., & Russo, D., (2007). Modelling the impact of agricultural abandonment and wildfires on vertebrate diversity in Mediterranean Europe. *Landscape Ecology, 22,* 1461–1476.

Morrison, J. C., Sechrest, W., Dinerstein, E., Wilcove, D. S., & Lamoreux, J. F. (2007). Persistence of large mammal faunas as indicators of global human impacts. *Journal of Mammalogy, 88*(6), 1363–1380.

Mouillot, F., Ratte, J.-P., Joffre, R., Mouillot, D., & Rambal, S. (2005). Long-term forest dynamic after land abandonment in a fire prone Mediterranean landscape (central Corsica, France). *Landscape Ecology, 20*(1), 101–112.

Ostermann, O. P. 1998. The need for management of nature conservation sites designated under Natura 2000. *Journal of Applied Ecology, 35*(6), 968–973.

Pausas, J. G., & Bradstock, R. A. (2007). Fire persistence traits of plants along a productivity and disturbance gradient in mediterranean shrublands of south-east Australia. *Global Ecology and Biogeography, 16*(3), 330–340.

Pausas, J. G., & Fernández-Muñoz, S. (2012). Fire regime changes in the Western Mediterranean Basin: From fuel-limited to drought-driven fire regime. *Climatic Change, 110*(1–2), 215–226.

Pausas, J. G., & Ribeiro, E. (2013). The global fire-productivity relationship. *Global Ecology and Biogeographyl, 22*(6), 728–736.

Pausas, J. G., Keeley, J. E., & Verdú, M. (2006). Inferring differential evolutionary processes of plant persistence traits in Northern Hemisphere Mediterranean fire-prone ecosystems. *Journal of Ecology, 94*(1), 31–39.

Pereira, H. M., Navarro, L. M., & Martins, I. S. (2012). Global biodiversity change: The bad, the good, and the unknown. *Annual Review of Environment and Resources, 37*(1), 25–50.

Pfeiffer, M., Spessa, A., & Kaplan, J. O. (2013). A model for global biomass burning in preindustrial time: LPJ-LMfire (v1.0). *Geoscientific Model Development, 6*(3), 643–685.

Pinhasi, R., Fort, J., & Ammerman, A. J. (2005). Tracing the origin and spread of agriculture in Europe. *PLoS Biology, 3*(12), e410.

Poschlod, P., Kiefer, S., Tränkle, U., Fischer, S., & Bonn, S. (1998). Plant species richness in calcareous grasslands as affected by dispersability in space and time. *Applied Vegetation Science, 1*(1), 75–91.

Proença, V. (2009). *Galicio-Portuguese oak forest of Quercus robur and Quercus pyrenaica: Biodiversity patterns and forest response to fire*. Tese de Doutoramento. Lisboa: Faculdade de Ciências da Universidade de Lisboa.

Proença, V., & Pereira, H. M. (2010). Appendix 2: Mediterranean Forest (pp. 60-67). Leadley, P., Pereira, H.M., Alkemade, R., Fernandez-Manjarrés, J.F., Proença, V., Scharlemann, J.P.W., Walpole, M.J. (Eds.) *Biodiversity Scenarios: Projections of 21st century change in biodiversity and associated ecosystem services*. Secretariat of the Convention on Biological Diversity, Montreal. Technical Series no. 50, 132 pages.

Proença, V., & Pereira, H. M. (2013). Species-area models to assess biodiversity change in multi-habitat landscapes: The importance of species habitat affinity. *Basic and Applied Ecology, 14,* 102–114.

Proença, V. M., Pereira, H. M., & Vicente, L. (2008). Organismal complexity is an indicator of species existence value. Frontiers in. *Ecology and the Environment, 6*(6), 298–299.

Proença, V., Pereira, H. M., & Vicente, L. (2010). Resistance to wildfire and early regeneration in natural broadleaved forest and pine plantation. *Acta Oecologica, 36*(6), 626–633.

Rey Benayas, J. M., Martins, A., Nicolau, J. M., & Schulz, J. J. (2007). Abandonment of agricultural land: An overview of drivers and consequences. *CAB reviews: Perspectives in Agriculture, Veterinary Science, Nutrition and Natural Resources, 2*(57), 1–14.

Ruddiman, W. F. (2013). The Anthropocene. *Annual Review of Earth and Planetary Sciences, 41*(1), 1–24.

Russell-Smith, J., Murphy, B. P., Meyer, C. P., Cook, G. D., Maier, S., Edwards, A. C., Schatz, J., & Brocklehurst, P. (2009). Improving estimates of savanna burning emissions for greenhouse accounting in northern Australia: Limitations, challenges, applications. *International Journal of Wildland Fire, 18*(1), 1–18.

Sandom, C. J., Hughes, J., & Macdonald, D. W. (2013a). Rooting for rewilding: Quantifying wild boar's *Sus scrofa* rooting rate in the scottish highlands. *Restoration Ecology, 21*(3), 329–335.

Sandom, C., Donlan, C. J., Svenning, J. C., & Hansen, D. (2013b). Rewilding. In D. W. Mcdonald & K. J. Willis (Eds.), *Key Topics in Conservation Biology* 2 (pp. 430–451). Oxford: John Wiley & Sons.

Sandom, C. J., Ejrnæs, R., Hansen, M. D. D., & Svenning, J.-C. (2014). High herbivore density associated with vegetation diversity in interglacial ecosystems. *Proceedings of the National Academy of Sciences, 111*(11), 4162–4167.

Schley, L., Dufrêne, M., Krier, A., & Frantz, A.C. (2008). Patterns of crop damage by wild boar (*Sus scrofa*) in Luxembourg over a 10-year period. *European Journal of Wildlife Research, 54*(4), 589–599.

Seddon, P. J., Armstrong, D. P., & Maloney, R. F. (2007). Developing the science of reintroduction biology. *Conservation biology, 21*(2), 303–312.

Shlisky, A., Waugh, J., Gonzalez, P., Gonzalez, M., Manta, M., Santoso, H., Alvarado, E., Nuruddin, A. A., Rodríguez-Trejo, D. A., & Swaty, R. (2007). *Fire, ecosystems and people: Threats and strategies for global biodiversity conservation.* Arlington: The Nature Conservancy.

Smith, F. A., Lyons, S. K., Ernest, S. K. M., Jones, K. E., Kaufman, D. M., Dayan, T., Marquet, P. A., & Haskell, J. P. (2003). Body mass of late Quaternary mammals. *Ecology, 84,* 3402.

Svenning, J. C. (2002). A review of natural vegetation openness in north-western Europe. *Biological Conservation, 104*(2), 133–148.

Thonicke, K., Venevsky, S., Sitch, S., & Cramer, W. (2001). The role of fire disturbance for global vegetation dynamics: Coupling fire into a dynamic global vegetation model. *Global Ecology and Biogeography, 10*(6), 661–677.

Turner, M. G. (1998). Landscape ecology, living in a mosaic. In S. I. Dodson et al., (Eds.), *Ecology* (pp. 78–122). New York: Oxford University Press.

Van Andel, J., & Aronson, J., (2012). *Restoration ecology: The new frontier.* Chichester: Wiley-Blackwell.

Vera, F. W. M. (2000). *Grazing ecology and forest history.* CABI Publishing, Oxon, UK.

Vera, F. W. M. (2009). Large-scale nature development-The Oostvaardersplassen. *British Wildlife, 20*(5), 28–36.

Verburg, P. H., & Overmars, K. P. (2009). Combining top-down and bottom-up dynamics in land use modeling: Exploring the future of abandoned farmlands in Europe with the Dyna-CLUE model. *Landscape Ecology, 24*(9), 1167–1181.

Whelan, R. J. (1995). *The ecology of fire.* Cambridge University Press, Cambridge, UK.

Willis, K. J., & Birks, H. J. B. (2006). What is natural? The need for a long-term perspective in biodiversity conservation. *Science, 314*(5803), 1261–1265.

Wright, J. P., & Jones, C. G. (2006). The concept of organisms as ecosystem engineers ten years on: Progress, limitations, and challenges. *BioScience, 56*(3), 203–209.

Zeder, M. A. (2008). Domestication and early agriculture in the Mediterranean Basin: Origins, diffusion, and impact. *Proceedings of the National Academy of Sciences, 105*(33), 11597–11604.

Zimov, S. A. (2005). Pleistocene park: Return of the mammoth's ecosystem. *Science, 308*(5723), 796–798.

Part III
Rewillding in Practice

Chapter 9
Rewilding Europe: A New Strategy for an Old Continent

Wouter Helmer, Deli Saavedra, Magnus Sylvén and Frans Schepers

Abstract The European landscape is changing and new opportunities for conservation are arising. The main driver of this change is an unprecedented shift in agricultural practices that started in the mid-twentieth century. As a result, shepherds and small-scale farmers release nearly 1 million ha of land from agriculture, each year. Although land abandonment is often seen as a major socio-economic problem, it could also be an opportunity for a new rural development based on nature and wild values. This idea can be further enhanced by the comeback of a number of iconic wildlife species, by an increased network of protected areas, by better legislation and enforcement, and a more favourable environment policy. Rewilding Europe responds to these major changes in the European landscape by ceasing this opportunity for both the European natural heritage and Europeans. The initiative aims to rewild 1 million ha of land by 2022, creating ten wildlife and wilderness areas all across Europe. Besides the ecological benefits of rewilding abandoned landscapes, wild values can create new opportunities for entrepreneurship in these areas, while a restored and preserved wildlife will attract many visitors to watch, enjoy and experience the wild. Ultimately, a large-scale shift in land use across Europe towards wilder nature and innovative ways to use this resource for employment and subsistence could be achieved, thus turning threats and problems into opportunities.

Keywords Rewilding Europe · Land abandonment · Wildlife comeback · Rewilding enterprise · Wild values · Key species

W. Helmer (✉) · D. Saavedra · M. Sylvén · F. Schepers
Rewilding Europe, Toernooiveld 1, 6525 Nijmegen, The Netherlands
e-mail: wouter.helmer@rewildingeurope.com

D. Saavedra
e-mail: deli.saavedra@rewildingeurope.com

M. Sylvén
e-mail: magnus.sylven@rewildingeurope.com

F. Schepers
e-mail: frans.schepers@rewildingeurope.com

H. M. Pereira, L. M. Navarro (eds.), *Rewilding European Landscapes,*
DOI 10.1007/978-3-319-12039-3_9, © The Author(s) 2015

9.1   The Opportunity of Change

Europe is changing rapidly, offering more opportunities for nature today than for the past centuries. One of the major reasons of this is an unprecedented change in land-use, a unique circumstance driven by three major forces: a strong migration of—in particular younger—people to the cities[1] (EC 2008), intensification of agricultural use on the most productive soils (e.g. Pinto-Correia and Mascarenhas 1999) and, at the same time, large scale land abandonment in more remote areas (Keenleyside and Tucker 2010). Each year nearly 1 million ha of land are abandoned by shepherds and small farmers. Where land abandonment is often seen as a major socio-economic problem, it may provide an opportunity for new forms of rural development based on nature and certain valuable attributes of wild landscapes (see Chap. 1).

This opportunity is complemented by the major comeback of a number of iconic wildlife species (Deinet et al. 2013; Enserink and Vogel 2006; Kuemmerle et al. 2010; Russo 2006; and see Chaps. 4, 5, 8), supported by a growing network of protected areas (especially Natura 2000) better designed to suit multi-use criteria, with, for example, strict conservation, development, and ecotourism (e.g. Geneletti and van Duren, 2008; Zhang et al. 2013); better legislation and enforcement (Habitats, Bird and Water Directives, Bern Convention); and a more favourable policy environment (Wilderness Resolution, new EU Biodiversity Strategy); all contributing to an historic opportunity to create more space for wild nature in Europe. By reacting to these developments, European conservationists can make significant steps forward in their efforts to create a robust network of ecosystems that can sustain and improve their ecological values based on natural processes. The main challenges to the conservation of Europe's natural heritage are not so much related to *where* and *what* to protect, but *how* to protect and manage these often considerably large areas, and to optimise their ecological potential.

A fundamental challenge to this process exists in reinforcing the relevance, importance and value of these vast natural areas to European citizens and both urban and rural communities (Hochtl et al. 2005; Lupp et al. 2011). Over the last 10 years a growing number of initiatives all over Europe are focusing on natural processes and the reintroduction of missing keystone species (e.g. Burton 2011; Decker et al. 2010; Sandom et al. 2013)[2] as a key conservation approach, as opposed to active human management. Because most of these, often stand-alone, projects focus on the broader trends described above, the need to combine the approach and create an opportunity for collaboration has emerged. This is now available through the European wide initiative: Rewilding Europe (Sylven et al. 2010).

---

[1] http://www.eea.europa.eu/pressroom/speeches/urbanisation-in-europe-limits-to-spatial-growth.

[2] *For more examples, see:*http://www.wildeurope.org/index.php?option=com_content&view=article &catid=2%3Arestoration&id=62%3Aconference-pres&Itemid=19.

9.2   A New Vision for an Old Continent

The Initiative

In November 2010, Rewilding Europe, a new European nature initiative was launched in Brussels. To jump-start the initiative, the initiating partners, WWF Netherlands, ARK Nature, Wild Wonders of Europe and Conservation Capital agreed only a few months later to establish the Foundation Rewilding Europe as a separate legal entity. Together with the foundation, a limited company was registered, fully owned by the foundation. In this way Rewilding Europe is able to set up innovative conservation enterprises and participate in new forms of sustainable business development related to rewilding activities.

Rewilding Europe aims to rewild 1 million ha of land by 2020 (Sylven et al. 2010), creating ten magnificent wildlife and wilderness areas to serve as inspirational examples for what can be replicated and achieved elsewhere. These ten areas should serve as benchmarks for a large-scale shift in land use across Europe towards wilder nature and new ways to use that resource for employment and self-sustainability (Schepers and Widstrand 2012). To support this, a wider European Rewilding Network is under development[3] with the ultimate aim of influencing land use over a total of 10 million ha.

Nominations from all over Europe

In May 2009, the first ideas for a new European nature initiative were presented to a wide audience at the first European Wilderness Conference in Prague (Coleman and Aykroyd 2009). Organizations, governments, park managers and relevant stakeholders were invited to nominate areas to potentially be part of the initiative. In total, nearly 30 nominations were received for areas with high rewilding potential from all corners of Europe.

Out of these, five prime regions were selected to become showcases of how the Rewilding Europe vision can be put into practice (Fig. 9.1). These areas are located in Western Iberia (Portugal and Spain), the Velebit Mountains (Croatia), the Eastern Carpathians (Slovakia and Poland), the Southern Carpathians (Romania), and the Danube Delta (Romania).

By incorporating the next four rewilding areas into the initiative (Fig. 9.1), a diverse geographical representation of Europe will be achieved covering a broad array of different landscapes, from lowland river deltas to high mountains; from dehesa to tundra; from primeval beech forests to taiga; from upland river valleys to high cliffs. Each of these areas covers a minimum potential size of 100,000 ha.

---

[3] http://www.rewildingeurope.com/news/articles/rewilding-europe-starts-european-rewilding-network/.

Current Rewilding Areas:
1. Western Iberia
2. Velebit Mountains
3. Eastern Carpathians
4. Southern Carpatians
5. Danube Delta

Candidate Areas:
6. Central Apennines
7. Greater Laponia
8. Odra Delta
9. Rhodopes

Fig. 9.1 Map of Europe showing the existing five rewilding areas (*purple*) and the four candidate areas (*grey*) in 2013

The rewilding areas are guided by three main principles: (1) Every area should host complete and naturally functioning ecosystems specific to the region with a full spectrum of native wildlife typical for the region present; (2) The areas should be embedded within the social, historical and cultural fabric of their respective region; and (3) The new land use should be based on what nature can offer and be economically viable and competitive with other alternatives. These principles were defined in order to show that Europe can indeed deal in new ways with nature, within a modern society, that gives space for wild areas, wildlife and wilderness. It is about letting nature run more of its own business—and at the same time letting people create businesses, jobs and employment from it. These attributes and opportunities are communicated by the initiative across a broad spectrum of stakeholders from the European Commission to local landowners. This is a completely new conservation vision for Europe driving the ultimate goal: a wilder Europe in the twenty-first century.

Main Objectives

Rewilding Europe has set itself 10 clear objectives to be achieved over a 10-year period. These are applied from a central governance to local level, and include:

1. A total of at least 1 million ha (10,000 km²) of land will be 'rewilded' by the initiative and its partners, across 10 places covering different geographical regions of Europe, including different landscapes and habitats.
2. A substantial wildlife comeback (in particular for keystone or flagship species) will take place in the 10 rewilding areas, supported by re-introductions where appropriate or necessary, serving as the starting point for complete, functional ecosystems.
3. In each of the 10 rewilding areas, sufficient "in-situ" breeding facilities for wildlife will be established, for a variety of wildlife species that can be used for re-introductions or re-stocking of these areas.
4. Because of a growing demand for wildlife in these rewilding areas, European wildlife will develop a 'market value', providing new business opportunities—for management partners, landholders, hunting associations and the like.
5. In each of the 10 rewilding areas, 'rewilding' will become a competitive form of land (and sea) use; through supporting and building of rewilding enterprises, the economic prospects of local people and/or communities will be improved.
6. Magnification of success: the 10 rewilding areas serve as inspiring examples for other areas in Europe. This should ideally lead up to 100 other 'rewilding' initiatives launched across Europe affecting a total of 10 million ha (100,000 km²).
7. "Wild nature & natural processes" will be accepted and adopted as one of the main management principles for nature conservation in Europe, in particular in the larger landscapes that have a conservation status (especially the wilder, large Natura 2000 areas).
8. Through the work of Rewilding Europe, and the communication & outreach thereof, a sense of 'Pride of the Wild' will be created among a very broad audience in Europe, who will also again be able to enjoy these wild values.
9. A science-based and practical, tailor-made monitoring system will be established to oversee progress on the objectives of Rewilding Europe, both at the central level and in the rewilding areas.
10. The concept of the 'Joy of the Wild' will have reached out to at least 350 million European citizens, using different kind of media, outdoor and indoor exhibitions, computer and mobile applications, etc.

The Operating Model

The Rewilding Europe operating model is centred on the 10 rewilding areas. There are three main components in this operating framework (Fig. 9.2).

OPERATING MODEL – REWILDING EUROPE

Fig. 9.2 Operating model of Rewilding Europe

The rewilding areas are in themselves carefully selected based on a number of criteria that together determine their critical success factors. Each rewilding area works in an integrated way on the three components, which are rewilding, enterprise development and communication. At the centre of this selection process are local rewilding partners who are critically important to drive and implement activities in the rewilding areas.

A central team devises the Rewilding Europe strategy and supports local teams to implement rewilding activities in the rewilding areas. The central team also launches tools and mechanisms to support programme activities, while addressing the three components listed above in an integrated way. For example it is possible that Rewilding Europe Capital provides a loan to a promising enterprise that is linked to a tauros (*Bos taurus*) breeding centre, while the animals are provided by the European Wildlife Bank (Fig. 9.2). The release of the animals is communicated to a wider European audience, in combination with the Aurochs book (Goderie et al. 2013) that describes the comeback of this European iconic species.

External partners and stakeholders provide support in various ways and are critical for Rewilding Europe's success and delivery. Among the strategic stakeholders are the **initiating partners** that provide strategic and technical support. **Financial partners** and funding institutions (some being also strategic partners) provide finance, such as the United Postcode Lotteries, Adessium Foundation, Liberty Wildlife Fund and new, future target groups such as impact investors and (local) business partners. **Local landholders and stakeholders**, such as private landowners, park and reserve managers, hunting concession owners and other landholders facilitate in securing land tenure and access. Finally **scientific institutions and experts** (both at a central and local level) provide scientific knowledge and background, do applied research and provide monitoring services. For example, together with Wageningen University, an international Wilderness Entrepreneurship Programme has started (see Chap. 10). Moreover, the Zoological Society of

London, BirdLife International and other local experts are currently undertaking feasibility studies and research work (e.g. Deinet et al. 2013).

This operating model is an evolving dynamic that adapts as lessons are learnt and the landscapes evolve, however it provides an overview of how the different activities and components of Rewilding Europe are interlinked and centred around the rewilding areas.

9.3 Applying the Model to the Rewilding Areas

General

By the end of 2011, all five rewilding teams were contracted and began working within Rewilding Europe. Naturally, the context of setting up rewilding projects is different in each of the localities based on the socio-economic situation, local policies, and the presence of local organizations that have the capacity to provide the right leadership.

The central Rewilding Europe team provided the necessary technical support to each one of the projects. Area visits were undertaken to work with the teams and to help the projects move forward. Input was given and experience shared on all subjects related to the objectives, from bison (*Bison bonasus*) reintroductions, archeozoology, wildlife watching, hide construction, business plans for nature tour outfitters, land tenure alternatives (such as community conservation areas and landowner agreements), and ways to find common ground with hunting interests.

Although differing from area to area, a good start has been made and the first achievements are encouraging. It is important to emphasize that rewilding is not a quick fix, it is not about going for short-term results only. A long-term commitment and support is required in which it is essential to carefully build a good understanding and base for rewilding, generate local support for the ideas, establish partnerships with local stakeholders and build up the momentum.

Rewilding

Preparatory work in all areas involves studies, mapping, local meetings, research, opinion surveys and other activities, to create the necessary base for future success. Each of the rewilding areas started by selecting pilot sites within their larger areas, that will become the starting points for the concrete rewilding and enterprise development on the ground.

The cooperation with certain government institutions that are key for rewilding (mainly Forest and Conservation Departments) has turned out to be challenging at times, because of traditional views, frequently driven by hunting, forestry or very intense traditional/subsidized management practice interests. Although governments are not expected to be among the first early adopters of the rewilding concept, their role is, of course, critical for either enabling or supporting the pilot projects in the rewilding areas. By 2014, the local teams had managed to sign agreements, Letters of Intent and Memorandums of Understanding (MoUs) with several crucially important government institutions.

Another (expected) challenge for the rewilding concept has been certain traditional misunderstandings about wilderness ecology in Europe, in which the crucial ecological role of wildlife has been largely underestimated, especially when it comes to the role of the large herbivores (but see Chap. 8). In order to increase wildlife numbers, a core part of the initiative is to try to create 'breeding zones' within the rewilding areas where hunting is prohibited and wildlife numbers left to be naturally determined. In several areas good progress has been made to start working with the local hunting communities, in particular on the Portuguese side of Western Iberia and in Velebit.

Feasibility studies for reintroductions have been done or are underway for beaver (*Castor fiber*), red deer (*Cervus elaphus*), and fallow deer (*Dama dama*) in the Danube delta, European bison (*Bison bonasus*) in the Southern Carpathians and Velebit, for Balkan chamois (*Rupicapra rupicapra*) in Velebit, and red deer, roe deer (*Capreolus capreolus*), and Spanish ibex (*Capra pyrenaica*) in Western Iberia. These studies are required by law to permit re-stocking or re-introduction activities.

Box 9.1: Major Rewilding Initiatives at the Central Level

Wildlife Recovery Programme: a major element of the rewilding component that works with experts from all over Europe is to support natural wildlife numbers in all the rewilding areas, through planning and preparing releases or reintroductions of targeted wildlife species, in particular European bison (Vlasakker 2014), red deer, wild horse (*Equus ferus caballus*) (Linnartz, L & Meissner, R. 2014), wild bovines, beaver, Iberian lynx (*Lynx pardinus*), Spanish ibex, chamois and others (started in 2011). This process is implemented through a number of pioneering initiatives.

European Wildlife Bank: a live asset-lending model designed to reintroduce and expand naturally grazing wild herbivore populations across Europe. This is set up as a rewilding enterprise support initiative together with ARK Nature, a partner organisation. The EWB focuses on large wild herbivores (started early 2013).

European Bison Rewilding Action Plan: a strategic action plan to create viable, wild bison populations of at least 100 individuals each in five of the rewilding areas by 2022. These animals will be sourced from existing populations e.g. zoos, nature reserves and private collections (operational in 2013).

European Rewilding Network: a network of smaller initiatives and areas in Europe where rewilding is a key target. These will be identified in addition to and in parallel with the 10 main focus rewilding areas. There are many dozens of other important initiatives over many countries, which need to be showcased and communicated about as they represent other pieces of the rewilding puzzle. Connected through this network, these areas will serve as examples to exchange experiences and learn from each other. This European Rewilding Network will be strongly influential in order to create a real European movement working for rewilding all over the continent (started in 2013).

The Tauros programme: an initiative to breed back an animal that as closely as possible resembles the original wild bovine species that once roamed all across Europe, and re-introduce them as wild functional species in European ecosystems. The aurochs went extinct in 1627, but its DNA is still very much alive, although spread between a few primitive cattle breeds which are now used for the Tauros programme. The goal is to breed significant numbers of these animals (for the time called Tauros) that can start to live in free and social herds in at least five rewilding areas by 2020. This is done in partnership with the Taurus Foundation.

Communication

Good progress was made in promotion for Rewilding Europe as a new initiative, to several of the main target audiences. Social media, an AV-show trailer and printed materials were developed. Following the initial publishing of the major GEO cover story "Europa Wird Wilder[4]" in Germany, a further 11 special country editions published the story in 2012. Feature articles were published in leading newspapers all over Europe and in 'The New Yorker'. Swedish, Dutch, Spanish and Portuguese national TV reported from the initiative and the first activities in the rewilding areas. Major outreach was created through the outdoor exhibitions in European capitals of which the one in Madrid reached around 104 million people through mass media coverage in the Spanish-speaking world. As a result of this, "rewilding" is now a widely understood phrase used more and more in all various contexts and translated into several languages.

A communication training was organized in the Netherlands in spring 2012, which was attended by all the local rewilding teams. At that level great achievements have been made, especially in Western Iberia, Velebit and the Eastern Carpathians, with the local teams trying out several different methods that range from opinion surveys, magazine publishing, film making, local fairs, local stakeholder meetings as well as meetings with ministers and serious media PR work around the outdoor exhibitions.

Professional photo assignments have been carried out in the five first rewilding areas, including documentations of the first animal releases done in Western Iberia and Southern Carpathians. This means that there is now a good stock of high quality imagery available from these areas that is widely used for all kinds of communication and promotion. The quality of the image resources and the way this footage is used are a significant part of what makes the initiative stand out from the crowd.

---

[4] "Europe is wild" in German.

The development of tangible local visions for each rewilding area was given a high priority. Part of that focus led to the start of the production of artist's impressions of the envisioned future landscapes, one for each area.

Enterprise Development

The enterprise development work has made good progress in preparing fertile ground for the development of businesses, both at the central level and the rewilding area level (with a focus on Western Iberia and Velebit).

The enterprise team has worked to develop a rewilding business-financing instrument called Rewilding Europe Capital (REC), a revolving financing facility funded by philanthropic capital and owned by Rewilding Europe. REC's initial goal is to provide small, but often crucially important loans to promising businesses in the rewilding areas (starting 2013) that can generate meaningful rewilding outputs. Furthermore a business proposal has been developed for a European Safari Company. In cooperation with experienced travel organizations, this company will develop a network of seasonal camps, supported by wildlife hides, fly camps, and a range of nature and wildlife based guided activities.

Development of relevant businesses goes hand in hand with the rewilding work (see Chap. 10). In both Western Iberia and Velebit extensive explorations were done to identify existing and potential businesses that could potentially leverage rewilding outputs (directly and indirectly). This work produced a list of more than 40 potential businesses and business ideas to support and help develop. A shortlist of the most promising and meaningful businesses was selected to engage further, including tourism lodges, adventure trails, tented camps, tourism operators and relevant hunting operators.

9.4   First Results in the Rewilding Areas

Although Rewilding Europe is only 3 years on the way, the results in the first five rewilding areas are quite promising (Table 9.1).

Western Iberia: Ancient Dehesa and Montado Landscapes

The Iberian Peninsula, with some of the earliest human settlements in Europe, is home to some of the most ancient cultural landscapes of the continent. One typical example is the Spanish "Dehesa" or the Portuguese "Montado", traditional wood pastures which date back to the middle ages (Fig. 9.3). The savannah-like appearance shaped by large grazers, especially cattle, is today home to some of the rarest animal species of Europe, such as the Spanish Imperial Eagle and the Iberian Lynx,

Table 9.1 Preliminary results in the first five rewilding areas for the 2012–2013 period, summarizing the main results per area, regarding the three components of the operating model: rewilding, communication and enterprise development

	Rewilding	Communication	Enterprise development
Western Iberia	Land stewardship agreements on more than 6000 ha signed	Three new wildlife watching hides	Network of local entrepreneurs set up
	Purchase of 200 ha in Portugal	Rewilding seminar organized	Agreements on advertising with 10 tour operators
	Feasibility study on red deer, roe deer and ibex	Rewilding brochure prepared	Building of guesthouse (8 beds) at Campanarios
	Release of first herds of horses and tauros	Biological station/visitor center completed in Campanarios	Support of a B&B on the Portuguese side
			Business proposition: Faia Brava camp
Eastern Carpathians	Study on natural grazing by wild-living horses completed	Film documentary about 'the wolf mountains'	Set up of an ecotourism and wildlife travel agency
	Campaign to halt the killing of wolves	Production of "Bieszczadnik" magazine	Selection and training of nature guides
		Establishment of walking trails with wildlife watching sites	Development of ecotourism packages
		Launch of educational geo-cache trail	
		Public opinion survey on rewilding and wilderness protection	
Velebit Mountains	Study on ecological role bark beetle completed	Seminar with local entrepreneurs on wildlife watching, breeding and no-hunting zones	Ministry of tourism supports wildlife watching as a key economic activity
	Archaeozoological study completed as input for reintroduction missing key species	Preparation of an outdoor exhibition in Zadar	Support of a lodge, near the bison breeding center
	Freshwater study completed		
	Preparation of breeding centers for bison, tauros and free living horses		

Table 9.1 (continued)

	Rewilding	Communication	Enterprise development
Southern Carpathians	Preliminary inventory of pristine forests in Tarcu mountains and request for their protection	Establishment of a technical wilderness working group with the directors and biologists of protected areas	Meeting with local entrepreneurs to explore first ideas for business opportunities
	Feasibility study of re-introduction of European bison carried out and preparations for a breeding center in Tarcu mountains		
	Guidelines for management of forest reserve areas with bark beetle outbreaks		
	First agreement on a small no-hunting zone around the bison area		
Danube Delta	Feasibility studies for the reintroduction of beaver (finished) and deer (draft)	HD video footage of the delta collected	Inventory of the existing local businesses
	MoU with Danube Delta Biosphere Reserve Authority (DDBRA) about rewilding	Underwater photographic mission executed	Preparations to set up a community conservancy

the most endangered feline in the world. These species have together with their favourite prey, the European rabbit (*Oryctolagus cuniculus*), decreased alarmingly in numbers during the last century and just only recently begun to come back slowly (Deinet et al. 2013).

Western Iberia is a bit of a frontrunner at this stage. This is mainly due to the fact that the two partner NGOs (Fundación Naturaleza y Hombre in Spain and Associação Transumança Natureza in Portugal) owned land approximates 1300 ha, which make up the Campanarios de Azaba and Faia Brava nature reserves. These two areas are pilot sites from which the rewilding process is starting and taking shape.

Both NGOs are adopting emerging rewilding concepts and activities transitioning from traditional ways of subsidized biodiversity conservation, over to rewilding approaches and rewilding enterprise development. For example in the reserves trees were previously being planted and now large herbivores are being introduced to stimulate the natural vegetation development.

Fig. 9.3 Dehesa/montado landscape in Western Iberia. (Photo credit: Staffan Widstrand/Rewilding Europe)

Regarding the signed land stewardship agreements, 2852 ha are with direct management rights for the partner NGO's and 3471 ha without. With the release of tens of primitive Retuerta and Garrano horses and with Sayaguesa and Maronesa cattle as part of a Tauros breeding programme, natural grazing has started in the two reserves. A network of local entrepreneurs is set up to become part of the "European Safari Company" in association with an international wildlife/nature tour operator.

Velebit Mountains: The Wild West of the Adriatic Coast

Velebit is situated on the Adriatic coast of Croatia. This limestone mountain chain is 145 km long from north to south, and lies parallel to the coast (Fig. 9.4). Following a cross section from the crystal waters of the Adriatic in the west, it rapidly rises to 1757 m, and then phases out into a higher-level plateau towards the east. The area hosts an extraordinary diversity of different habitats, from barren Mediterranean landscapes at sea level, a large network of spectacular caves, to almost boreal systems at higher altitudes. This has led to the establishment of the two Paklenica & Northern Velebit National Parks as well as the Velebit Nature Park. Together the three areas occupy more than 220,000 ha. The area is declared a UNESCO Man and Biosphere Reserve and has been included in the UNESCO Tentative List of World Heritage Sites. Inside the Nature Park and outside the protected areas in the

Fig. 9.4 Limestone peaks in the Velebit mountains. (Photo credit: Staffan Widstrand/Rewilding Europe)

south and west, there are very promising areas for rewilding, consisting mainly of abandoned farmland and grazing lands.

Lobby and advocacy at government level for rewilding, by the local partner WWF in Croatia, has generated genuine interest, however due to changes in government composition and policies, this interest is difficult to consolidate (e.g. revisions in hunting legislation are postponed by the parliament). Due to very active networking on the ground, a lot of promising contacts with local entrepreneurs and hunting associations have been made.

The study on bark beetle (*Ips typographus*) emphasizes the importance of this species in opening up forests as a key natural process in the area. The archaeozoological study on the historical presence of larger mammals, proved among others the existence of ibex (*Capra ibex*) in Velebit. Because the availability of water is a key limiting factor in these limestone mountains, a freshwater study was executed to map water resources for establishing natural wildlife densities in the area. An overview of the (19) existing hunting concessions has identified opportunities for creating large breeding zones and negotiations have started with several concessionaires. Finally a successful seminar was held with local entrepreneurs focusing on wildlife breeding in two places and wildlife watching in combination with the creation of breeding sites.

Eastern Carpathians: One of Europe's Top Wildlife Areas

The Eastern Carpathians border an area between Poland and Slovakia, forming one of the wildest corners of Europe including vast, extensive forests with untamed rivers, low undulating mountains with scattered alpine meadows, and pockets of old-growth forests (Fig. 9.5). Here, one of Europe's largest wild-living populations of bison lives side by side with red deer, roe deer, wild boar (*Sus scrofa*), lynx (*Lynx lynx*), wolves (*Canis lupus*), bears (*Ursus arctos*), beavers, and otters (*Lutra lutra*). Few other regions of the continent have more protected areas than the Eastern Carpathians—in total around half a million ha of national parks, biosphere reserves, forest reserves, landscape parks, nature parks and Natura 2000 sites. However there is still a lot to improve on the protection of old growth forests, natural wildlife numbers and the development of a wilderness based economy.

A feasibility study shows that the Eastern Carpathians rewilding area provides huge rewilding opportunities on both the Slovakian and Polish side. This mainly focuses on trans-boundary wilderness management of migratory species including large herbivores and carnivores between the two countries. However, the level of commitment from key local stakeholders to work on rewilding is still unclear today. The first year of the project was used to create a base for a rewilding perspective with the general public. A public opinion survey on both sides of the border showed major support for the concept of wilderness protection and rewilding, thereby creating new economic opportunities.

Fig. 9.5 Extensive forest with untamed rivers—the San river in Eastern Carpathians. (Photo credit: Grzegorz Leśniewski/Wild Wonders of Europe)

The two local partner organizations (WOLF in Slovakia and the Carpathian Wildlife Foundation in Poland) put most of their energy in campaigns for rewilding and wilderness protection. Thus the starting up of pilots on the ground is lagging behind. WOLF significantly contributed to prevent an amendment to the Game Act, which would have extremely threatened herbivores and large carnivores in the Eastern Carpathians. They also achieved that selective trapping of carnivores will not be allowed. Though the legal limit in Slovakia is set to 130 killings of wolfs per year, 150 were killed in 2012, a third of which in the rewilding area. The local partner, WOLF, has been running a campaign to halt these killings, using petitions sent directly to the European Commission.

Southern Carpathians: A Wilderness Arc at the Heart of Europe

At the southern end of the Carpathian Mountains in Romania, an initiative is underway to create one of Europe's largest wilderness landscapes south of the Arctic Circle. With a backbone of more than 1 million ha of protected areas already in place, large intact forests, a high concentration of biodiversity, un-fragmented landscapes, wild rivers, and large mosaic landscapes still kept open by small scale farming practices, there is a unique opportunity to realise this vision.

The starting point is an area around the Tarcu Mountains Natura 2000 Site, with connections to the Domogled-Valea Cernei National Park, and the Retezat National Park, which together cover around 200,000 ha. The area includes a wide variety of ecosystems—alpine meadows and grasslands, old beech and fir forests, steep cliff formations, and undulating mosaic landscapes with open grasslands intersected by woodlands (Fig. 9.6). However, the numbers of large carnivores and herbivores are depressingly low and, due to poisoning, most scavengers and all vultures are gone. Patches of virgin forest are still threatened by illegal logging.

During the first year, the Southern Carpathians has concentrated on stakeholder meetings, feasibility studies, GIS mapping and planning, and creating a support base for the rewilding concept in the project area through a series of meetings with local people.

The rewilding area is part of the (larger) South Western Carpathians Wilderness Area project[5], run by WWF Romania and covering 11 Protected Areas and the present rewilding area itself, which is promising for further expansion over time of the rewilding activities.

A preliminary inventory of pristine forests areas has been performed in the Tarcu Mountains N2000 area and a request was made to decision makers for their protection. An agreement was reached with the Romanian government on criteria for identification and inclusion of pristine forests in stricter protection status (Ministerial Order). The guidelines for management of forest reserve areas with bark beetle

---

[5] http://www.erweiterungsbeitrag.admin.ch/en/Home/Projekte/Projekt_Detailansicht?projectinfo ID=222831.

Fig. 9.6 Alpine meadows and old growth forests in the Southern Carpathians. (Photo credit: Staffan Widstrand/Rewilding Europe)

outbreaks were developed and submitted to the Romanian government, to promote as a key natural process and non-intervention policy. Managers of hunting areas, game and forest managers were approached about initiating non-intervention management measures (e.g. creation of no-hunting areas). Finally a feasibility study of re-introduction of European bison was carried out, with the most suitable area identified and agreed for re-introduction in the Tarcu Mountain Natura 2000 site, to be executed in 2014.

Danube Delta: Europe's Unrivalled Wetland

The Danube Delta on the border between Romania and Ukraine is outstanding in Europe—due to its size (over 600,000 ha), intact river dynamics, unexploited coastline (shaped by the Danube River and the Black Sea together), wide horizons and large-scale landscapes without significant infrastructure (Fig. 9.7). It also has the largest reed beds in the world, in addition to millions of nesting and migrating birds, many of them rare and some even globally endangered. However, some of the key wildlife species are still missing, such as wolf, red deer and beaver. Due to poisoning, numbers of species like black kites (*Milvus migrans*), golden jackals (*Canis aureus*) or vultures are extremely low. Apart from legal enforcement, giving value to these species as part of the wildlife watching economy should stimulate social control to avoid these illegal practices.

Fig. 9.7 The Danube delta, Europe's largest delta. (Photo credit: Staffan Widstrand/Rewilding Europe)

The unique Letea Forest mosaic savannah, situated in the Romanian section, is one of the few "primeval" forests of the country that has trees up to 700 years old. Through the designation as UNESCO Biosphere Reserves by both the Romanian and Ukrainian governments, with some relatively strictly protected core areas, the delta enjoys a high level of formal protection. Buffer areas and economic zones around these also provide opportunities for local developments without jeopardizing the natural values.

A lot of effort was spent the first year to ensure that the rewilding project is firmly anchored in two main communities in the outer and drier part of the delta, and with the relevant authorities at several levels. A key concept that should be tried in the Danube Delta is the development of one or two community wildlife conservancies[6], alongside with reintroduction of species, and with wildlife tourism providing income to these communities. This will allow for developing several different rewilding enterprises that start providing jobs and income to the people who live here.

To achieve this, geographical boundaries of the possible conservancies in Letea and Sfantu Georghe are identified in collaboration with local stakeholders. An inventory was made of the existing local businesses and other operating businesses in Sfantu Georghe, which is important information for the establishment of the community conservancy. Regular meetings for setting up a Community Conservancy in Sfantu Georghe and CA Rossetti municipalities brought mutual trust in and

---

[6] For an example, see the program of the Namibian Association of Community Based Natural Resource Management (http://www.nacso.org.na/).

knowledge of the Rewilding Europe initiative. An MoU has been signed between the Danube Delta Biosphere Reserve Authority (DDBRA) and the local rewilding partner, WWF Romania, about the development of the rewilding area. Finally feasibility studies for the reintroduction of beaver and deer are subcontracted to ICAS, a research centre at the Brasov Wildlife Department. The deer release is in preparation.

9.5 A Future Outlook for Rewilding Europe

Rewilding Europe has presented a compelling vision about the historic opportunities that Europe is facing, and how we could make this a reality. With a media-outreach of more than 100 million people in the two first years, the initiative seems to capture the imagination of many Europeans (Schepers and Widstrand 2012). Support was received from all strands of society: local communities and governments, landowners, hunters, scientists, NGOs, EU Parliamentarians, local entrepreneurs and top business people. Practical work is starting: the first releases of key species, wildlife tourism developments, and small legal achievements. Nonetheless, most of the work is still in a stage of studying or negotiation with stakeholders. Between vision and practice there is a lot to do, and which needs a lot of support. The coming years will prove if Rewilding Europe can bridge the gap between vision and practice. 'Making it real' is therefore the slogan for the coming years.

Key for a successful continuation of the programme is a prosperous start of the several large scale projects that are on the way, such as the first European community conservancy in the Danube Delta (26,000 ha) and some agreements on better hunting practices in Velebit and Western Iberia, on a scale of tens of thousands of hectares, proving that the scale that Rewilding Europe is pursuing is not unrealistic. Just as important is a successful start up of some serious rewilding enterprises with a consistent spin-off, such as the European Safari Company and related enterprises in the rewilding areas, showing that an alternative rural economy can really be build in abandoned areas. Furthermore a careful selection of the next four pilot areas (completing the 'Rewilding 10 of Europe' objective in 2014) will help the initiative to illustrate that opportunities for rewilding exist in every corner of Europe.

Finally it's crucial to build on a strong relation between the local teams that do most of the work, and a central team that facilitates them in their rewilding, communication and enterprise activities. Rewilding Europe believes that real change can only come from ownership and leadership of those organizations and entities that nominated their areas to become part of the wider initiative.

References

Burton, A. (2011). Where the wisents roam. *Frontiers in Ecology and the Environment, 9*(2), 140–140.

Coleman, A., & Aykroyd, T. (2009). *Wild Europe and large natural habitat areas.* (Conference Proceedings). Prague, Czech Republic.

Decker, S. E., Bath, A. J., Simms, A., Lindner, U., & Reisinger, E. (2010). The return of the king or bringing snails to the garden? The human dimensions of a proposed restoration of European Bison (*Bison bonasus*) in Germany. *Restoration Ecology, 18*(1), 41–51. doi:10.1111/j.1526–100X.2008.00467.x.

Deinet, S., Ieronymidou, C., McRae, L., Burfield, I. J., Foppen, R. P., Collen, B., & Bohm, M. (2013). *Wildlife comeback in Europe: The recovery of selected mammal and bird species.* Final report to Rewilding Europe by ZSL, BirdLife International and the European Bird Census Council, London, UK.

EC. (2008). *Poverty and social exclusion in rural areas* (p. 243). European Commission—DG employment, social affairs and equal opportunities.

Enserink, M., & Vogel, G. (2006). The carnivore comeback. *Science, 314*(5800), 746.

Geneletti, D., & van Duren, I. (2008). Protected area zoning for conservation and use: A combination of spatial multicriteria and multiobjective evaluation. *Landscape and Urban Planning, 85*(2), 97–110.

Goderie, R., Helmer, W., Kerkdijk-Otten, H., & Widstrand, S. (2013). *The aurochs, born to be wild* (p. 160). The Netherlands: Roodbont.

Hochtl, F., Lehringer, S., & Konold, W. (2005). "Wilderness": What it means when it becomes a reality—a case study from the southwestern Alps. *Landscape and Urban Planning, 70*(1–2), 85–95. doi:10.1016/j.landurbplan.2003.10.006.

Keenleyside, C., & Tucker, G. (2010). *Farmland abandonment in the EU: An assessment of trends and prospects.* WWF Netherlands and IEEP.

Kuemmerle, T., Perzanowski, K., Chaskovskyy, O., Ostapowicz, K., Halada, L., Bashta, A. T., & Radeloff, V. C. (2010). European bison habitat in the Carpathian mountains. *Biological Conservation, 143*(4), 908–916.

Linnartz, L., & Meissner, R. (2014). Rewilding horses in Europe. Background and guidelines—a living document. Rewilding Europe, The Netherlands.

Lupp, G., Höchtl, F., & Wende, W. (2011). "Wilderness"—A designation for central European landscapes? *Land Use Policy, 28*(3), 594–603.

Pinto-Correia, T., & Mascarenhas, J. (1999). Contribution to the extensification/intensification debate: New trends in the Portuguese montado. *Landscape and Urban Planning, 46*(1–3), 125–131.

Russo, D. (2006). *Effects of land abandonment on animal species in Europe: Conservation and management implications.* Italy: UniversitÀ degli Studi de Napoli Federico, Napoli.

Sandom, C. J., Hughes, J., & Macdonald, D. W. (2013). Rooting for Rewilding: Quantifying wild boar's sus scrofa rooting rate in the Scottish highlands. *Restoration Ecology, 21*(3), 329–335. doi:10.1111/j.1526–100X.2012.00904.x.

Schepers, F., & Widstrand, S. (2012). *Rewilding Europe: Annual review* (p. 60). The Netherlands: Rewilding Europe.

Sylven, M., Wijnberg, B., Schepers, F., & Teunissen, T. (2010). *Rewilding Europe—Bringing the variety of life back to Europe's abandoned lands* (p. 30). WWF.

Vlasakker, J. van de (2014). Bison Rewilding Plan 2014–2024. Rewilding Europe's contribution to the comeback of the European bison. The Netherlands.

Zhang, Z., Sherman, R., Yang, Z., Wu, R., Wang, W., Yin, M., & Ou, X. (2013). Integrating a participatory process with a GIS-based multi-criteria decision analysis for protected area zoning in China. *Journal for Nature Conservation, 21*(4), 225–240.

Chapter 10
Preparing a New Generation of Wilderness Entrepreneurs

Lessons from the Erasmus Intensive Programme 'European Wilderness Entrepreneurship' 2013

Judith C. Jobse, Loes Witteveen, Judith Santegoets and Derk Jan Stobbelaar

Abstract This chapter discusses the role of education in the preparation of the next generation of entrepreneurs in nature conservation. Departing from the traditional conservation education, which emphasizes ecological management, the chapter is a plea for incorporating entrepreneurship in the curricula of educational programmes on rewilding ecosystems. An Erasmus Intensive Programme on European Wilderness Entrepreneurship is presented as a case study. A set of competences is defined and operationalized based on the evaluation of the first edition of the programme undertaken in Rewilding Europe's pilot area in Western Iberia. Aspects of the learning strategies and learning environment are presented and reviewed. The conclusion of this chapter is that to learn wilderness entrepreneurship competences, an environment should be created in which students, teachers and stakeholder co-learn at the boundaries of their comfort zones.

Keywords Entrepreneurship · Education · Competences · Learning strategies · Erasmus Intensive Programme · Western Iberia

J. C. Jobse (✉) · J. Santegoets · D. J. Stobbelaar
Van Hall Larenstein University of Applied Sciences
Velp, The Netherlands
e-mail: judith.jobse@wur.nl

J. Santegoets
e-mail: judith.santegoets@wur.nl

D. J. Stobbelaar
e-mail: derk-jan.stobbelaar@wur.nl

L. Witteveen
Van Hall Larenstein University of Applied Sciences
Wageningen, The Netherlands
e-mail: loes.witteveen@wur.nl

H. M. Pereira, L. M. Navarro (eds.), *Rewilding European Landscapes,*
DOI 10.1007/978-3-319-12039-3_10, © The Author(s) 2015

10.1   Introduction

Ever since the creation of national parks and other large-scale nature conservation efforts, the principal objectives of conservation, science, and recreation are bound together in an ever-changing interrelated triangle. Their dependencies, benefits and tensions are in a constant flow, often defined by financial contexts (Kupper 2009). This is also discernible in the current European situation of dwindling subsidies and uncertain finances for nature conservation, which calls for conservation activities that generate revenues to achieve economically sustainable conservation. Several actors in European wilderness conservation are adopting conservation strategies that aim to achieve economic sustainability. For instance, Rewilding Europe included wilderness-based entrepreneurship in their main objectives (see Chap. 9), and the NGO Wild Europe (wildeurope.org) launched an economic benefits group in 2013 to stimulate "a new breed of wilderness warrior". Due to the rise of these new European wide conservation initiatives there is a need to train students and professionals in wilderness entrepreneurship.

Nature conservationists are traditionally not trained in entrepreneurship and business development as it is often assumed that professions related to nature conservation are principally guided by ecological and sustainable principles. Those fields therefore define the educational design of nature conservation curricula. This chapter describes an exploration into this new field of education in the context of rewilding European abandoned agricultural land.

10.2   Entrepreneurship in Conservation Education

We found that scholars in conservation science hardly mention entrepreneurship in their publications. Using entrepreneur* as a search term in JSTOR Data for Research (http://dfr.jstor.org), we found that the term was introduced around the 1920's and 105,555 articles have used the term by 10 July 2013. Most of these articles are classified by JSTOR under the business (31%), economics (26%), history (16%), political science (15%), and sociology (11%) disciplines. We found that scholars in biological sciences, ecology, plant sciences, and zoology do not often mention ($\leq 0.6\%$ of all articles per discipline) entrepreneurship in their publications (Fig. 10.1). When searching with the same term in a few major conservation journals in JSTOR on 15 July 2013, we found 48 (out of 4140) articles in Conservation Biology, 8 (out of 3149) in Ecological Applications, 1 (out of 14025) in Ecology, 3 (out of 4384) in Journal of Applied Ecology, and 4 (out of 9353) in The Journal of Wildlife Management.

Many nature conservationists have a degree in biological sciences or from a resource management programme. These programmes started to expose their students to the importance of other scientific fields due to the emergence of the new interdisciplinary field of conservation biology in the 1980's (Meffe and Carrol 1997). Although, economics and sociology are recognised as important fields within conservation biology, entrepreneurship is not mentioned by Meffe and Carrol (1997).

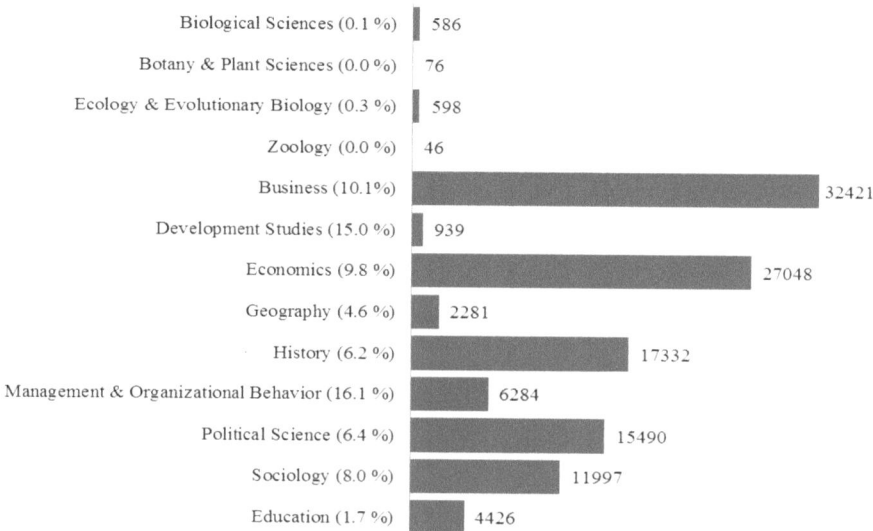

Fig. 10.1 Number of journal articles found when searched for entrepreneur* per discipline on JSTOR Data for Research (http://dfr.jstor.org) on 10 July 2013. Percentages are calculated by dividing the number of articles found by the total number of articles published per discipline. Individual journals can be included in more than one discipline

Nor is it included in the recommended guidelines for conservation literacy from the Education Committee of the Society for Conservation Biology (Trombulak 2004).

The European Commission (2006) recognises the importance of entrepreneurship in (higher) education and mentions entrepreneurship as one of the eight key competences for lifelong learning. Nonetheless, they reported that "the teaching of entrepreneurship is not yet sufficiently integrated in European higher education institutions' curricula" (European Commission 2008). They further conclude that the majority of entrepreneurship courses are offered in business and economic studies.

Since 2000, Dutch governmental policies are in place to stimulate entrepreneurship in Dutch education (Van der Aa et al. 2012). This policy encouraged several Dutch universities of applied sciences to integrate entrepreneurship in their curricula. Currently, entrepreneurship is part of curricula of several programmes in agribusiness, forestry and nature management, agriculture and rural development at the authors' home institution.

10.3   Case Study: The Erasmus Intensive Programme on European Wilderness Entrepreneurship

In the project 'European Nature Entrepreneur', the NGO Rewilding Europe and three Dutch educational institutes who deliver life sciences, rural development and nature education at vocational, bachelor, master and PhD levels, collaborate to develop new curricula that innovate on thematic contents and corresponding learning strategies. In this project, new curricula are being developed to offer (nature) entrepreneurship competences for students with an interest in forestry and nature conservation, wildlife management, applied animal sciences, rural development, and sustainable tourism/recreation. The project envisages students of different educational levels collaborating with each other and with professionals in Rewilding Europe's pilot areas.

One of the project outputs is a 14-day Erasmus Intensive Programme on European Wilderness Entrepreneurship (IP EWE) funded by the European Lifelong Learning Programme. IP EWE is an experimental course that introduces the concept of wilderness entrepreneurship to students from various disciplines in higher education in Europe. The first edition of this IP was held in Rewilding Europe's pilot area in Western Iberia in spring 2013.

The Western Iberian pilot area was proposed to Rewilding Europe by a Portuguese and a Spanish NGO, Associação Transumância e Natureza (ATN) and Fundación Naturaleza y Hombre (FNYH), respectively. Officially, since the end of 2011, Rewilding Europe and the local NGOs collaborate to develop and execute a strategy for rewilding an area of 100,000 ha (see Chap. 9). Within the pilot area, the two local NGOs own and manage two nature reserves (Faia Brava and Campanarios de Azaba) and they are in the process of buying more land to expand their reserves. Buying land is not an easy task even when money becomes available (from donations and commercial activities), because owners in this region are often absent or not known due to a lack of registrations in the Portuguese national cadastre. The small sizes of properties makes it even more complicated in the Portuguese region; Faia Brava includes 860 ha of land, which is divided in 140 different properties (Beukers 2013). Besides managing their own land, the NGOs also make agreements with other landowners to manage their land for conservation purposes. For instance, ATN owned in 2013 around 650 of the 860 ha of the Faia Brava reserve and the remaining land is owned by others. Some owners still use their land to grow olives, almonds or grapes, while others only harvest cork or do not use their land at all. Scaling up to rewild 100,000 ha is a real challenge and can only take place if local stakeholders (landowners and governments) support the idea, even if farmland continues to be abandoned. Just north of Faia Brava there is even an expansion of vineyards due to rising exports of Douro wines. Besides the reserves managed by the two NGOs, there are other natural protected areas in the region recognized by the Portuguese and Spanish authorities. None of these areas is designated as wilderness, but they are mostly managed to protect certain species and scenic landscapes. The Western Iberian pilot was an interesting location for the IP because both ATN and FNYH work together with Rewilding Europe to explore and experiment with wilderness entrepreneurship in and around their reserves (see Chap. 9).

During the IP, a group of thirty students and fifteen lecturers from seven higher education institutes (HEIs) in Bulgaria, Croatia, the Netherlands, Portugal, Spain and Sweden looked at the role of entrepreneurship to promote the future wilderness of Western Iberia. Participating students were enrolled in bachelor or master programmes in Biology, Agriculture, Tourism, Sustainable Land Planning, Land

Fig. 10.2 Execution of landscape observation exercise using creativity tools to train students to look from different perspectives in Celorico da Beira, Portugal. (Photo Credit: Judith Jobse)

Management, Environmental Science, Forestry and Nature Conservation, Tropical Forestry, Landscape Architecture, Socio-spatial Analysis, and Communication.

The IP was organised in a variety of locations in Spain and Portugal: Aveiro (Portugal), Ciudad Rodrigo (Spain) and Figueira de Castelo Rodrigo (Portugal). During the first day at the University of Aveiro, the programme was explained and students were introduced to each other. In Aveiro, and also later in the programme at other locations, lectures were given on subjects such as rural tourism, wildlife management, planning and land use strategies, but also about transdisciplinarity and the actor network theory. During the second day the IP group moved to the Mediterranean landscape of Rewilding Europe's pilot region in Western Iberia. On the way to the region, participants were given a landscape observation exercise using creativity tools (Fig. 10.2). The tools (sketchbook, drawing pencil, 3 colour pencils, eraser, and pencil sharpener) were later used in an exercise to create a group vision for an area, and students used these tools more spontaneously when preparing for their final presentations.

In the pilot region, participants visited a variety of natural and cultural attractions in the pilot area such as the Spanish and the Portuguese nature reserves owned by the NGOs, historical sites, such as the ruins and the restored village of Castelo Rodrigo, the Côa Valley Archaeological Park and the Côa Museum. They experienced the low population density in most of the small villages in the region where mostly retirees were encountered. Meetings with various local stakeholders were organized to discuss the problems and possible alternatives for regional development and nature conservation. Among stakeholders were: the NGOs responsible for the local nature reserves and Rewilding Europe; mayors of Portuguese and Spanish villages; the president of a hunting association, local entrepreneurs such as producers and retailers of cheese, jam, almonds, wine, and Iberian pig meat products; and the owners of a bar and a bed & breakfast.

The main assignment for the IP students was to explore economic dynamics that could contribute to the ecological restoration and future wilderness in the region. In groups of three to four, they articulated a vision for the region, together with a business model that would give both the ecological system and the local community new perspectives. Many of the business models that students produced were in the tourism sector, which is in line with what Rewilding Europe promotes in this region (see Chap. 9). Some student groups perceived the lack of publicity for the area as its main problem. Therefore, they came up with a web portal for local (tourism) enterprises and a marketing brand for the whole region. They also came up with models to expand the local variety of tourist activities. One group proposed to diversify the local economy by introducing a snail farm from which part of the profits would go to nature conservation and education. Two groups proposed ideas to stimulate the use of land, which would perhaps only fit with Rewilding Europe's vision for the area if these activities were outside of a core wilderness area. One of these groups launched the idea for an organization that could bring conservation volunteers from all over the world to Western Iberia to help maintain the agricultural production on some of the terraces. Another group proposed a company that mediates between local communities and businesses new to the area to smoothen the purchase of land. A final business model introduced the concept of small self-sustaining office units in the landscape to rent to people who would like work in a sustainable office with a great view for inspiration. During a 'market presentation', which was open to the public, students shared their visions and business models with local stakeholders and interested community members (Fig. 10.3). One of the IP students arranged during the IP her return to the region to conduct a MSc thesis research project on the use of social learning to increase levels of local involvement (Leuvenink 2013). Another MSc student helped organize the IP as an internship assignment and stayed in the region after the IP to conduct an analysis of the relation between Rewilding Europe, the Portuguese conservation partner ATN, and the local population (Walet 2014).

10.4 Designing a Wilderness Entrepreneurship Curriculum

The process of putting together a new IP with seven different universities spread out over Europe is a complicated task. It was the first time that this consortium of universities worked together to offer a collaborative curriculum and organizing the programme became a priority, rather than the documentation of the educational de-

Fig. 10.3 Locals inhabitants of Figueira de Castelo Rodrigo (Portugal) listen to students presenting their vision for the region and their business model. (Photo Credit: Judith Jobse)

sign process. Even though articulated programme principles were not documented beforehand, it is possible to carry out a critical analyses of the IP EWE.

The case study is based on observations of participating lecturers, on the minutes of a review meeting and on the results of a questionnaire distributed to participants at the end of the programme. Students participated in the whole IP. From the 30 participants, 20 returned the questionnaire including 9 bachelor and 11 master students (5 Bulgarians, 2 Croatians, 6 Dutch, 2 Portuguese and 5 Spanish). Most lecturers (11 out of 15) participated in more than 50% of the programme and 8 returned the questionnaire (2 Bulgarians, 4 Dutch, 1 Spanish and 1 Swedish). Nine lecturers joined a review meeting at the end of the programme.

Creating new entrepreneurship curricula or integrating entrepreneurship in existing curricula requires the identification of the competences that students should attain. In the Netherlands, competence based learning is mainstream at universities of applied sciences. Competences encompass knowledge, skills, and attitude. They "enable successful task performance and problem solving with respect to real-world problems" (Lans et al. 2013). A second aspect to consider is the way students learn these competences; which learning strategies are used for the different competences? When describing those learning strategies, we include all learning methodologies applied such as lectures and practical work and the role of teachers and stakeholders in the process. A third aspect to consider is the environment in which learning takes place. Learning environments have certain qualities that enable or disable the specific learning strategy that is envisaged. We divide the learning environment into physical and social aspects. For the social learning environment, we focus on cultural and linguistic aspects. The physical learning environment often used in formal education is a classroom setting. Literature shows that changing this setting—getting outdoor, working in other cultures or countries, in new landscapes—can increase the learning capacity of students (Meijles and Van Hoven 2010; Peacock and

Pratt 2011; Nedovic and Morrissey 2013). These new or unfamiliar environments can increase motivation, enhance imagination, and create focus.

Wilderness Entrepreneurship Competences

For the first edition of IP EWE, the competences for entrepreneurship were not clearly articulated before the programme implementation. However afterwards, a set of competences that capture the learning goals of the IP were identified. Based on a free interpretation of the study by Lans et al. (2013), and in line with the initial ideas that shaped the IP, five clusters of competences for wilderness entrepreneurship were constructed:

- Opportunity competence, which refers to problem spotting, an eye for innovation, a sense of creativity and foresight thinking. It is also considered as an action oriented competence with aspects of self-efficacy.
- Social competence, which refers to dealing with diversity, interdisciplinarity and multi-stakeholder contexts at the interpersonal level referring to communication, facilitation and enabling participation.
- Normative competence, which refers to the ability to deal simultaneously with diverse dimensions. These dimensions may be perceived as conflictive yet require to be integrated in a sustainability perspective such as economic, ethic, political social and environmental dimensions. This integrative view on society and environment makes that this competency also deals with moral decision-making and citizenship (Closs 2011).
- Complexity competence, which refers to the ability to focus on complex problems and system thinking. It is the ability to identify and analyse (sub) systems and domains, and the ability to understand and reflect on boundaries and interdependencies. This competence also refers to the ability to explore uncertain futures (Enserink 2010).
- Business competence, which refers to the ability to realise and manage project and business opportunities.

The main IP assignment aimed for the integration of the above-mentioned competences by giving the IP students the task to create a business model for regional development in Western Iberia. This assignment challenged students, while working in interdisciplinary and international teams, to transfer problems into business opportunities and to come up with innovative and sustainable solutions in a complex situation within multi-stakeholder contexts.

Students operationalized the business competence on the level of creating a business model while the other aspects of the business competence (ability to realise and manage a business) was considered outside the scope of this IP. When creating a business model students also worked on the opportunity competence as it deals with innovation and creativity. From the questionnaire it became clear that students recognized this competence: "*I learned how to make my idea creative and at the same time convincing.*" Students remarked that being creative is hard, but considered it a must in solving new kinds of problems and added that an open mind is necessary. When asked what they learned about nature entrepreneurship, the majority of the participants indicate that they now have a broader view on the topic. They now see that entrepreneurship is a nature conservation strategy that brings opportunities and have a reservoir of examples on how to connect business to nature. Especially the students with a background in ecology or ecology-related subject considered nature entrepreneurship as an eye opener.

During this assignment, students were deliberately put together in mixed groups. The interdisciplinary and international groups led students to work on their social competence. They had to deal with diversity and interdisciplinarity and came to learn the importance of social skills such as negotiating and clearly communicating. The majority of the participants indicate that this interdisciplinarity was very important and needed for good results or right decisions. Students reflect on this as follows: "*Interdisciplinarity is necessary to make the right decisions when solving complex situations*". They also express doubts: "*Not sure if it worked, because it felt more as if we were just adding different disciplines together (so including everything) instead of going in between.*" Learning outcomes in relation to intercultural communication and language were mentioned by many students: "*I learned that intercultural communication is even harder than I remembered and that it is challenging to stay open minded while you feel like others don't.*" In a general way they praised the collaboration with all the different cultures, because it opened up new ways of thinking about and dealing with the issues involved: "*Different cultures also leads to different interpretations of issues. Learning how to deal with this can minimize the conflicts. It was really important.*"

During the stakeholder meetings, students developed their social, normative, and complexity competences. They came to learn the importance of social skills such as communication while recognising the complexity of a situation in which different norms come together. A student remarked that the Rewilding Europe's concepts sounded pretty easy and logical in theory, but that is was almost impossible to do in practice. Although recognized as hard, students appreciated being exposed to the diverse opinions as this quote exemplifies: "*One thing can be bad for someone and good for another person. It is good to hear different and sometimes even completely opposed opinions about one thing.*"

Learning outcomes regarding the potential of the Rewilding Europe initiative and entrepreneurship further laid the foundation for the complexity and normative competences. Whereas students had seen the potential of the Rewilding Europe initiative before and used to consider it as a 'simple' answer to land abandonment, they increasingly realised the associated complexities and the diverse contexts. They were able to contest the concept of 'wilderness' and realised the relevance of local support. This made them doubt the possibilities of rewilding areas. Normativity became also apparent when dealing with the concept of nature entrepreneurship. Students remarked that they changed their view on 'making money' and started to see it as a necessity for a sustainable company or NGOs. This concern for the economic aspects seems to be a delicate issue as one student concludes that this is especially important for those conservationists or ecologists who seem to think that "nature is more important than mankind and that money is wrong": "*You can also use money to do good things.*"

The majority of the participants considered networking and the use of networks as a very important part of entrepreneurship, which relates to both the social and the business competences. Students reflected on that aspect with remarks about the importance of networking for the development of business. They acknowledge especially the examples provided. Networking is mostly regarded as positive but students also problematized it: "*Networking is very important to make choices that work for a longer time, but it also makes things more difficult, because opinions of people differ. I was thrown between different world views when speaking to one person and then to another.*"

Learning Strategies for Wilderness Entrepreneurship Education

During the IP, a large variety of learning methodologies were used, such as lectures, meetings with local people and stakeholders, field visits, group work, scenario simulation, roleplaying and informal conversations. When participants were asked what made learning interesting and effective, most participants pointed out the variety of people and perspectives. This was expressed in excursions, open discussions and group work and not so very much in lectures. Most of the participants favour the "untraditional" approach when explaining interesting and effective learning. Participants gave some critique on the lecturing activities, which were sometimes copies of formal academic lectures. Even though for most lectures the location was not in a school, the set-up was similar to formal educational settings. Lecturers suggested more interactive lectures and discussions, to prevent long days with traditional lectures.

Both students and lecturers considered that the programme would gain from more reflection and enhanced connections between the various activities. They noted that they would like the programme to be less or differently intensive, but recognised at the same time that this intensiveness was important for not getting distracted. It created a sense of connection with the local issues and an atmosphere where innovation could take place.

During the programme several moments were built in for students to engage with local stakeholders. These engagements consisted mostly of students receiving information from the stakeholders to get them acquainted with the local situation. The information flow between students and stakeholders was reversed during the final presentations of the students' business models to which the local community was invited (Fig. 10.3). The media attention for the IP, with broadcasts on several Portuguese television stations and dissemination on various websites, can be labelled as another form of stakeholder interaction. IP participants also recommended diversifying engagement with the local community, as this quote from a lecturer exemplifies: "*To amplify our cross-pollination I would add more informal gatherings, particular with youngsters, using the school and the teachers as gate-keepers.*"

Learning Environments for Wilderness Entrepreneurship Education

The physical learning environment during the IP was the Rewilding Europe pilot area in Western Iberia, as it was assumed that being present in an area where conservation NGOs actively experiment with wilderness entrepreneurship would enhance the learning process. It gave participants the chance to observe and experience the landscape, and to communicate with the stakeholders involved. Students praised this aspect, expressing that the fact that they '*were there*' was of major importance. Nonetheless, lecturers stated that the programme, the area, and the stakeholders require sound introduction. As a matter of fact, they observed that '*being there*' and '*talking to stakeholders*' did not automatically lead to a good understanding of the situation. Although field visits and excursions occupied a considerable part of the programme, a significant amount of time was spent between walls for lectures, workshops and presentations. This provoked some critique especially from lecturers as they assumed that the IP provided a very conducive outdoor environment with unique opportunities which are most appreciated by students. The regular indoor confinement was considered as a sub-optimal use of the available learning environment.

The IP offered a rich social learning environment (causing students to work on their social competences) because of the variety of learning methodologies applied, and the fact that participating students and lecturers were from many different nationalities and educational backgrounds. Another aspect of the social learning environment was the importance of a common language, which was English in this case. A student explained: "*I learned that it is hard to communicate when not all the people can speak English. Especially when you have to work with each other.*" Difficulties were most significant during the group work of students. They experienced limited language proficiency as the inability to express oneself clearly in English or working with someone who is not able to do so. Lecturers indicated that considering the language issues, time is needed for setting the scene, respecting cultural and contextual requirements, as well as considering the diversity of roles that define the learning experience.

10.5 Lessons Learned for Wilderness Entrepreneurship Education

By articulating the lessons learned during the IP, we aim to contribute to a growing understanding of educational programme design to prepare the next generation of wilderness entrepreneurs in Europe.

Our first lesson originates from evaluating competences in the IP. The partition of competences as described can be contested, as they seem to ignore overlapping elements. For example, it can be argued that consulting stakeholders is an essential aspect of both social and complexity competences. Stakeholder consultation deals with communication in practice and with appreciating diverse, and even contrasting, views on problem perceptions and alternatives. The formulated competences can equally be contested for ignoring important requirements of (learning) wilderness entrepreneurship. For example, the aspect of critical reflection is not articulated in any of the competences. Similarly absent is the aspect of conflict transformation, which is often mentioned as part of interaction between local people and nature conservationists (Martin 2012). However, the five formulated competences provide a good entry point to design education on wilderness entrepreneurship, taking into account that the competences are interlinked and overlapping. This leads to a first lesson learned: curricula for wilderness entrepreneurship should include the following competences: opportunity competence, social competence, normative competence, complexity competence and business competence.

Creativity cannot be learned from textbooks or by thematic lecturing, nor can other important aspects of wilderness entrepreneurship such as negotiating or dealing with complexity. These are learned best when they are put into practice. Therefore, it might be supportive to frame the field visits as action research to move away from its understanding as an outdoor lecture. Considering local informants and stakeholders as partners in the learning experience enhances the relevance for all parties involved. Such an approach articulates opportunities for a more circular knowledge exchange between students, teachers and stakeholders, which could be indicated as a social learning process. Reed et al. (2010) defines social learning as "a change in understanding that goes beyond the individual to become situated within wider social units or communities of practice through social interactions between actors within social networks". These thoughts lead to formulating a second lesson learned: the learning strategy for wilderness entrepreneurship should be all inclusive. All actors involved have to meet and engage in exchange of knowledge, expertise, opinions and other communicative resources (see also Leistra & Stobbelaar 2015).

The second lesson requires that learning for wilderness entrepreneurship competences should take place where the action is happening. On top of that, we also found that these competences seem to flourish outside formal educational settings. Creativity and out of the box thinking, both essential for building a new future for wilderness entrepreneurship, need unique experiences in which learners embark on unknown activities with a basic feeling of comfort. This consideration of comfort zones coincides with Wals (2007) whom states "The trick is to learn on the edge of people's individual comfort zone with regards to dissonance: if the process takes places too far outside of this zone, dissonance will not be constructive and block learning. However, if the process takes place within peoples' comfort zones—as is the case when homogenous groups of like-minded people come together—learning is likely to be blocked as well". Other authors refer to dissonance as issues of friction and congruence between self-regulation and external regulation (Vermunt and Verloop 1999). When analysing these findings we realised it is not only learning on the edge but also 'instruction on the edge'. This analysis leads to the formulation of a third lesson learned: wilderness entrepreneurship takes place where the practice is discernible and aspects of dissonance should be added to this learning environment, such as intensiveness and exposure to different cultures, disciplines and backgrounds. The environment should challenge both learners and lecturers in a way that learning takes place at the boundaries of comfort zones, building on positive friction between self and external regulation.

The three lessons learned relate to each other and provide tools to use for curriculum development on wilderness entrepreneurship. To learn wilderness entrepreneurship competences, an environment should be created in which students, teachers and stakeholders learn from each other in a challenging way. Once educational designers find ways to get good programmes put in place, we may have filled the earlier identified gap between nature conservation curricula and the current European context, which calls for conservation activities that generate revenues to achieve economic sustainable conservation.

References

Beukers, V. (2013). The relation between land ownership/land use and nature management in Portugal. Setting up a new geographic database with Rewilding Europe's local partner Associação Transumância e Natureza. Internship report bachelor student, Velp (The Netherlands): Van Hall Larenstein University of Applied Sciences.

Closs, L., & Antonello, C. S. (2011). Transformative learning integrating critical reflection into management education. *Journal of Transformative Education, 9*(2), 63–88.

Enserink, B., Hermans, L., Kwakkel, J., Thissen, W., Koppenjan, J., & Bots, P. (2010). *Policy analysis of multi-actor systems*. The Hague: Lemma.

European Commission. (2006). Recommendation of the European parliament and of the council of 18 December 2006 on key competences for lifelong learning. *Official Journal of the European Union, 394*, 10–18.

European Commission. (2008). Entrepreneurship in higher education, especially within non-business studies. Final Report of the Expert Group on the promotion of SMES competitiveness Entrepreneurship Delivery Strategies of EE.

Kupper, P. (2009). Science and the national parks: A transatlantic perspective on the interwar years. *Environmental History, 14*(1), 58–81.

Lans, T., Blok, V., & Wesslink, R. (2013). Learning apart and together: Towards an integrated compentence framework for sustainable entrepreneurship in higher education. *Journal of Cleaner Production.* doi:10.1016/j.jclepro.2013.03.036.

Leistra, G. R., & Stobbelaar, D. J. (forthcoming in 2015). The challenges of a green UAS given the post-normal governance condition. Procedia - Social and Behavioral Sciences. Paper presented at 5th World Conference on Learning Teaching and Educational Leadership, Prague, Czech Republic, 29–30 October 2014.

Leuvenink, A. (2013). Facilitating social learning to increase levels of local involvement: the case of Associação Transumância e Natureza in Portugal. MSc thesis Communication Science. Wageningen (The Netherlands): Wageningen University. Retrieved on November 4, 2014 from http://edepot.wur.nl/303040

Martin, B. (2012). Global values, local politics: Inuit internationalism and the establishment of northern Yukon National Park. In S. Höhler, P. Kupper, & B. Gissibl (Eds.), *Civilizing nature: National parks in transnational historical perspective* (pp. 157–172). Oxford: Berghahn Books.

Meffe, G. K., & Carroll, C. R. (1997). *Principles of conservation biology.* Sunderland: Sinauer Associates Inc.

Meijles, E., & Van Hoven, B. (2010). Using the rural atelier as an educational method in landscape studies. *Journal of Geography in Higher Education, 34*(4), 541–560.

Nedovic, S., & Morrissey, A. M. (2013). Calm active and focused: Children's responses to an organic outdoor learning environment. *Learning Environments Research, 16*(2), 1–15.

Peacock, A., & Pratt, N. (2011). How young people respond to learning spaces outside school: A sociocultural perspective. *Learning Environments Research, 14*(1), 11–24.

Reed, M. S., Evely, A. C., Cundill, G., Fazey, I., Glass, J., Laing, A., Newig, J., Parrish, B., Prell, C., Raymond, C., & Stringer, L. (2010). What is social learning? *Ecology and Society.* 15(4):r1 [online]. Retrieved on November 4, 2014 from http://www.ecologyandsociety.org/vol15/iss4/resp1/

Trombulak, S. C., Omland, K. S., Robinson, J. A., Lusk, J. J., Fleischner, T. L., Brown, G., & Domroese, M. (2004). Principles of conservation biology: Recommended guidelines for conservation literacy from the education committee of the society for conservation biology. *Conservation Biology, 18*(5), 1180–1190.

Van der Aa, R., Van Geel, S., & Van Nuland, E. (2012). *Ondernemerschap in het onderwijs. Tweemeting Eindrapport. Agentschap NL, namens Ministerie van OCW en Ministerie van EZ.* Rotterdam: ECORYS.

Vermunt, J. D., & Verloop, N. (1999). Congruence and friction between learning and teaching. Learning and instruction, *9*(3), 257–280.

Walet, L. A. (2014). Negotiating the production of space: the implementation of rewilding in North-East Portugal. MSc thesis Cultural Geography. Wageningen (The Netherlands): Wageningen University. Retrieved on November 4, 2014 from http://edepot.wur.nl/305191

Wals, A. E. (Ed.). (2007). *Social learning towards a sustainable world: Principles, perspectives, and praxis.* Wageningen: Academic Publishers.

Chapter 11
Towards a European Policy for Rewilding

Laetitia M. Navarro and Henrique M. Pereira

Abstract Millions of hectares of agricultural land could be released from human pressure within the next decades in Europe. Rewilding presents a great opportunity to restore the abandoned landscapes, along with the biodiversity and the supply of those ecosystem services that were until now restricted to the remaining few wild areas of the continent. As a result, rewilding is in a dire need of a policy framework in the European Union, to promote its implementation as a land management option, to evaluate its outcomes, and to share knowledge and good practices among stakeholders. In this chapter, we review the history of conservation policies and protected areas in the EU, the implementation of the Natura 2000 Network being one of the major milestones. We also discuss the role of conservation in sectoral activities such as agriculture. We present the growing importance given to wilderness areas and the inclusion of wilderness management into European policies. We then evaluate the contribution of wilderness and rewilding to the achievement of global and EU targets. Finally, recommendations are made to efficiently and adequately include rewilding into the European framework of conservation policies.

Keywords Nationally Designated Protected Areas · Natura 2000 · High Nature Value Farmland · Agri-Environmental Schemes · Wilderness · Conservation targets · Land management policies

L. M. Navarro (✉) · H. M. Pereira
German Centre for Integrative Biodiversity Research (iDiv) Halle-Jena-Leipzig
Deutscher Platz 5e, 04103 Leipzig, Germany
e-mail: laetitia.navarro@idiv.de

Institute of Biology, Martin Luther University Halle-Wittenberg
Am Kirchtor 1, 06108 Halle (Saale), Germany

Centro de Biologia Ambiental, Faculdade de Ciências da Universidade de Lisboa
Campo Grande, 1749-016 Lisboa, Portugal

H. M. Pereira, L. M. Navarro (eds.), *Rewilding European Landscapes,*
DOI 10.1007/978-3-319-12039-3_11, © The Author(s) 2015

11.1   Introduction: A Historical Perspective

Though evidence of land conservation goes back several thousands of years in Europe, the concept of protected areas was first implemented across the continent by the fifteenth century, when poaching and logging were banned from royal hunting forests by the nobility in order to protect the game (Jones-Walters and Čivić 2013; Possingham et al. 2006; Ramão et al. 2012). Those protected areas (PAs) were designed to preserve a given resource (e.g. timber or game), rather than to preserve nature in general. It was not until the nineteenth century that landscapes would be preserved for their "natural beauty", following a movement initiated in Germany to preserve *Naturedenkmal*, i.e. nature monuments (Jones-Walters and Čivić 2013). At the same time, the first "National Parks" (NP) were designated in the USA, in Yosemite NP, in 1864, then Yellowstone NP, in 1872 (Possingham et al. 2006), with the aim of preserving nature for recreational, cultural and ethical reasons (Borrini-Feyerabend et al. 2013). In 1909 the first European park was created in Sweden (Pinto and Partidário 2012; Ramão et al. 2012). Yet, it was not until the second half of the twentieth century that the official definition of "National Parks" was given by the IUCN as the first resolution of its 10th assembly (IUCN 1969).

The 1970s later mark a change in the way Protected Areas were managed in Europe, shifting from strict protection to the acknowledgment of the role and needs of local communities and other stakeholders, and their integration in the management of the landscape (Jones-Walters and Čivić 2013; Ramão et al. 2012). In 1971, the UNESCO launched the Man and Biosphere (MAB) program, leading to the concept of Biospheres Reserves in 1974 (Coetzer et al. 2014). It was followed, 20 years later, by the establishment of the World Network of Biosphere Reserves (UNESCO 1996), with a particular focus on the involvement of local communities, and their sustainable use of the resources present within the area. 1971 is also the year of the signature of the Ramsar Convention, for the global cooperation and conservation of wetland habitats (Possingham et al. 2006). In 1972, the UNESCO also signed the "Convention concerning the Protection of the World Cultural and Natural Heritage" (World Heritage Centre 2013). The first EU Natural Heritage Sites were established in 1979, in Croatia (Plitvice Lakes National Park) and in Poland (Białowieża Forest). In 2013, the EU28 counted 27 "natural" and 6 "mixed" Heritage sites (whc. unesco.org). More recently, wilderness areas have been given more importance in the EU, including with the acknowledgment of their role in biodiversity conservation (European Parliament 2009), while the abandonment of remote agricultural areas can be seen as an opportunity to increase the area of wild land via rewilding (see Chap. 1).

In this chapter, we first present the status and trends of current biodiversity conservation in the European Union, via the national designation of Protected Areas, the Natura 2000 network, and agri-environmental schemes. We then discuss the recent integration of wilderness in the EU conservation framework, along with the potential of rewilding abandoned farmland. We evaluate rewilding and wilderness conservation in regards to the achievement of the global and European conservation targets set for 2020. This chapter only discusses continental conservation and marine protected areas were removed from the analysis.

11.2   Current Conservation Policies in the EU

Nationally Designated Protected Areas

Nationally Designated Protected Areas (NDPAs) encompasses a variety of designations: "national park", "regional park", "nature park", "nature reserve", "biosphere reserve", "wilderness area", "wildlife management area", "landscape protected area", and "community conserved area" (Dudley et al. 2008; Ramão et al. 2012), which also vary greatly in their management policies. When countries are divided into "federal" states (e.g. Spain, Germany), each entity can also have regional designation policies. Moreover, some countries protect specific ecosystem nation-wide (e.g. wetlands in Croatia, rivers in Portugal), without designating them within their protected areas (Ramão et al. 2012). More than 31 % of the European NDPAs cover forest ecosystems, while agro-ecosystems are represented in over 28 % of the areas (Ramão et al. 2012). These areas also tend to be designated in mountain regions, due to their remoteness and the resulting lower human densities.

The IUCN defined, in 1994, six protection categories for the NDPAs (Dudley et al. 2008), based on the level of management and the allowed degree of human activity (Table 11.1), though not all areas are yet classified, or even registered as such. In practice, the managers of a given protected area report its protection category on a voluntary basis. Out of the 68 % of NDPAs classified by IUCN categories in Europe ($N = 52,995$), the vast majority belongs to category IV, Habitats/Species management areas (Table 11.1). However, category V (Protected landscape/seascape) covers the largest area on the continent. It is also interesting to observe that the strictest PAs in terms of management (Categories I and II) are not the most common, both in terms of number and area, with coverage of 20 % of the total protected areas. Nonetheless, although comparatively few areas are in category II (National Parks), they cover an area almost similar to the most represented type of protected area, category IV (respectively 88155 and 88352 km^2 in Table 11.1).

The historical distribution of the different types of NDPAs matches the history of the European perception of the role of protected areas. From the 1950s to the mid −1960s about half of the PAs were in the most restrictive categories (mostly national parks, Cat.II), while the other half were managed with the inclusion and/or tolerance of human activity (Fig. 11.1). In the 1970s there was a large increase of PAs designated as the less restrictive category V (Protected landscape). Currently, the IUCN categories II and I represent less than a quarter of the total classified PAs of Europe (Fig. 11.1).

Table 11.1 Description of the different IUCN categories for protected areas and contribution of the continental Nationally Designated Protected Areas of Europe to those categories. (Dudley et al. 2008; EEA 2013a)

Category	Name	Management type	Detail	Number (%)	Total area in km^2 (%)
Ia	Strict nature reserve[a]	Strict protection	Strictly protected areas set aside to protect biodiversity and also possibly geological/geomorphological features, where human visitation, use and impacts are strictly controlled and limited to ensure protection of the conservation values. Such protected areas can serve as indispensable reference areas for scientific research and monitoring	4514 (6%)	14549.18 (2%)
Ib	Wilderness area	Strict protection	Usually large unmodified or slightly modified areas, retaining their natural character and influence, without permanent or significant human habitation, which are protected and managed so as to preserve their natural condition	1207 (2%)	34672.43 (5%)
II	National park[b]	Ecosystem conservation and protection	Large natural or near natural areas set aside to protect large-scale ecological processes, along with the complement of species and ecosystems characteristic of the area, which also provide a foundation for environmentally and culturally compatible spiritual, scientific, educational, recreational and visitor opportunities	320 (<1%)	88155.57 (13%)
III	Natural monument or feature	Conservation of natural features	Protected areas set aside to protect a specific natural monument, which can be a landform, sea mount, submarine cavern, geological feature such as a cave or even a living feature such as an ancient grove. They are generally quite small protected areas and often have high visitor value	3124 (4%)	4571.65 (1%)

Table 11.1 (continued)

Category	Name	Management type	Detail	Number (%)	Total area in km^2 (%)
IV	Habitat/species management area	Conservation through active management	Protected areas aim to protect particular species or habitats and management reflects this priority. Many category IV protected areas will need regular, active interventions to address the requirements of particular species or to maintain habitats, but this is not a requirement of the category	31654 (41%)	88352.17 (13%)
V	Protected landscape/seascape	Landscape/seascape conservation and recreation	A protected area where the interaction of people and nature over time has produced an area of distinct character with significant ecological, biological, cultural and scenic value, and where safeguarding the integrity of this interaction is vital to protecting and sustaining the area and its associated nature conservation and other values	10837 (14%)	319117.34 (47%)
VI	Protected area with sustainable use of natural resources	Sustainable use of natural resources	Protected areas conserving ecosystems and habitats, together with associated cultural values and traditional natural resource management systems. They are generally large, with most of the area in a natural condition, where a proportion is under sustainable natural resource management and where low-level non-industrial use of natural resources compatible with nature conservation is seen as one of the main aims of the area	1339 (2%)	35044.49 (5%)
Not classified		N/A	N/A	24420 (32%)	97781.40 (14%)

[a] Two protected areas were assigned to category I, without distinction between Ia and Ib, and were not counted in this table

[b] Areas designated as "National parks" in Europe can fall in different IUCN categories than II

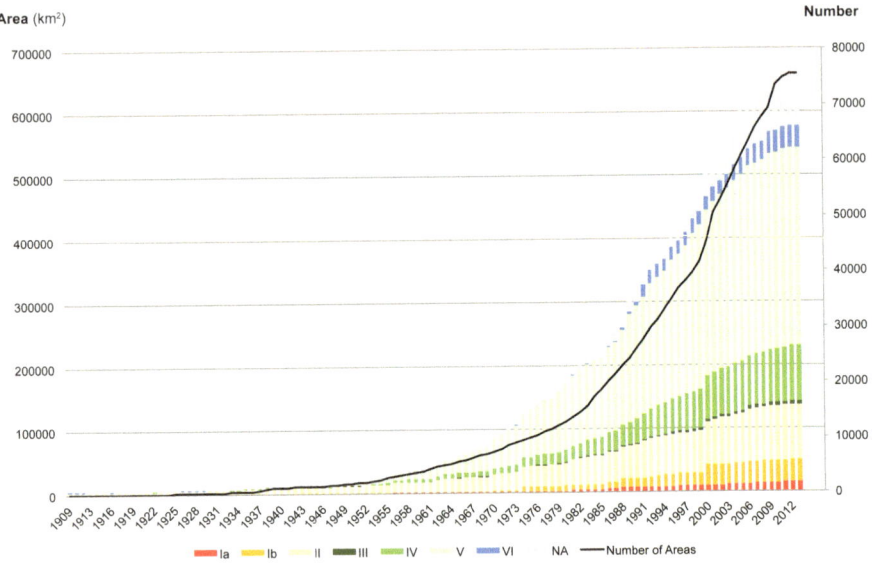

Fig. 11.1 Temporal evolution of the number of Nationally Designated Protected Areas in Europe, and the total area protected. The NDPAs are classified according to the IUCN categories, NA meaning that the area was not yet classified. (EEA 2013a)

Birds and Habitats Directives

The Council of Europe signed the Bern Convention in 1979 to give a legal framework for the conservation of biodiversity on the continent (Jones-Walters and Čivić 2013). This was followed by the adoption of the Birds Directive by the nine Member States of the EU in April 1979 (79/409/EC), to respond to worrying decreases in bird populations observed on the continent, and acknowledging that some species of birds are a European heritage and that the conservation of migratory species is a transboundary matter. The Directive was later amended, as new Member States joined the EU and was updated in November 2009 for the EU27 countries (Directive 2009/147/EC).

The directive's articles engage the Member States, *inter alia*, into maintaining populations at viable levels, creating protected areas and managing bird populations within and outside those areas. Particular attention should be given to bird species in Annex I (193 species), while species in Annex II (82 species) may be hunted under national legislations. The directive resulted in the creation of Special Protection Areas (SPAs), which number increased steadily, including with the addition of new Member States to the European Union.

The Birds Directive was followed, 13 years later, by the Habitats Directive (92/43/EEC), adopted in May 1992. This directive emphasizes the conservation of biodiversity via the conservation of "habitats, wild fauna and flora", in a context of sustainable development for the continent. A total of over 230 habitat types and over 1000 species of animals and plants were selected. Country specific lists of Sites of Community Importance (SCIs) were then evaluated by the Commission, before being implemented as Special Areas of Conservation (SACs) by the Member States (European Commission, 2002; Gaston et al. 2008a). The Habitats Directive further aimed at building a "coherent European ecological network", the Natura 2000 Network, which would encompass the Protected Areas created under the Birds Directive of 1979, the SPAs, and the newly designated SACs. The contribution of each EU country to the Natura 2000 network depends on the proportion of habitats (in annex I) and habitats for species (in annex II and IV) present within their borders.

The management of the Natura 2000 areas is the responsibility of each Member State, which can delegate and decentralize to federal or regional agencies (European Commission 2002). Traditional European landscapes may serve as a conservation baseline (Gaston et al. 2008a), as the guidelines on Natura 2000 site management emphasize the importance of ensuring "the continuation of traditional management regimes, which very often have been critical in creating and maintaining the habitats which are valued today" (European Commission 2002).

The Natura 2000 network is a unique example of a regional, transboundary, and unified network of protected areas (Crofts 2014; Hochkirch et al. 2013). As of 2008, Denmark and the Netherland had reached 100% of their sufficiency for the Habitats Directive Annex I habitats and Annex II species, meaning that their network of PAs covered at least one instance for 100% of the habitats and species of the annexes that had a known distribution on their territories (EEA 2009a). The rest of the EU Member States had between 70 and 99% of sufficiency, with the exception of Lithuania (61%), Czech Republic (59%), Cyprus (25%) and Poland (17%). The Natura 2000 Network is now shifting from establishing the areas to defining proper coordinated management strategies (European Commission 2013).

Overall Picture of Protected Areas in the EU

The ensemble of protected areas in the European Union, composed of Nationally Designated Protected Areas (NDPAs), Special Areas of Conservation (SACs), and Special Protection Areas (SPAs) is extensively covering the continent (Fig. 11.2a). As of 2013, the EU28 counted over 77,000 terrestrial NDPAs and nearly 23,000 continental Natura 2000 areas. Yet, 30% of the area protected in Europe represents an overlap between a type of designation or another. As a matter of fact, in some countries, such as Spain, Slovenia, and Estonia, the Natura 2000 areas almost entirely overlap with NDPAs (Fig. 11.2a). At the European scale, the overlap is particularly true for NDPAs in the IUCN categories I to IV (Ramão et al. 2012).

The majority of the Member States count more than 18% of their territories in a protected area (Fig. 11.2b). Nonetheless, the map of Europe depicts a different picture when focusing on the most restrictive conservation categories of the IUCN (Categories I and II on Fig. 11.2c): most countries have less than 3% of their area in those categories. Sweden, Belgium, and Slovakia are the only countries protecting

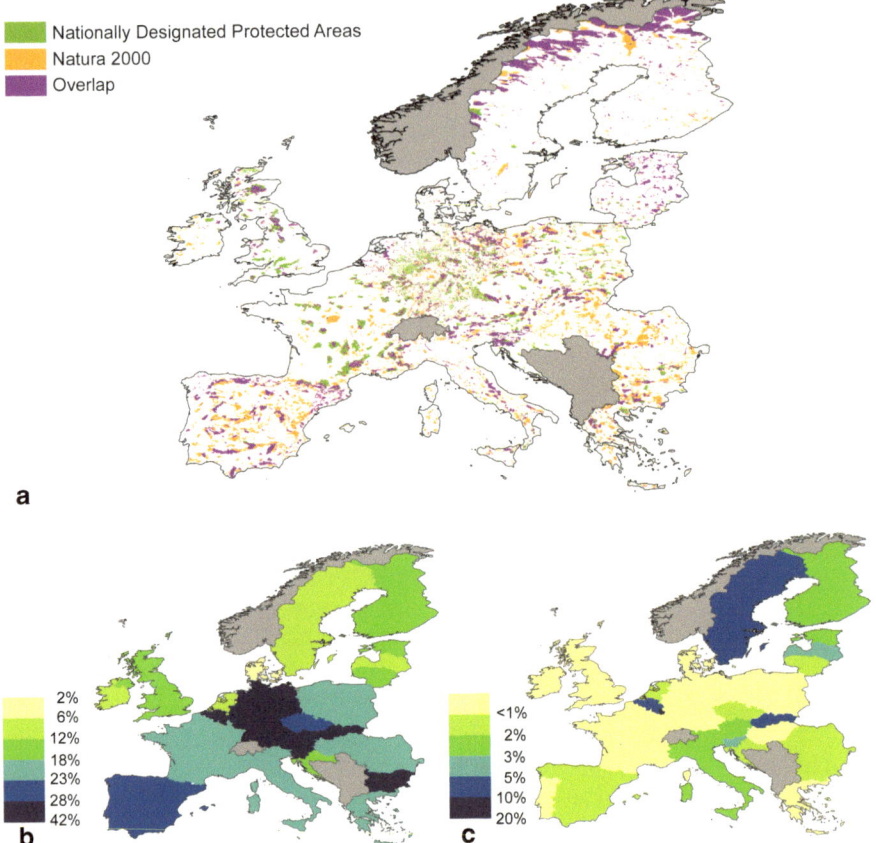

Fig. 11.2 Spatial perspective on European protected areas (EEA 2013a, 2013b). **a** European network of Nationally Designated Protected Areas, Natura 2000 sites, and overlap between the two
designations; **b** Proportion of each EU28 country within a protected area (Nationally Designated
Protected Areas and Natura 2000 sites); **c** Proportion of each EU28 country within a protected area
in category I or II of the IUCN

more than 5 % of their national area as a strict nature reserve, a wilderness area, or
a National Park (Fig. 11.2c). Natura 2000 areas overlapping with NDPAs classified
as categories Ia and Ib represent 4 % of the network (European Commission 2013).

The EU Protected areas tend to be created in high and remote areas, with lower
productivity (Dudley et al. 2008; Gaston et al. 2008b), and with less regard for the
habitats and the species that inhabit them than for the availability of the land. Nonetheless, conflicts might arise with local populations when an area used for resource
extraction is set to be strictly protected. Such tensions are exacerbated by strictly
top-down approaches, i.e. with the lack of consultation of local stakeholders in the
establishment of a PA, which is often the case with the establishment of Natura
2000 areas (Crofts 2014). On the contrary, less restrictive categories, or "multiple
use" PAs are typically more easily accepted (Possingham et al. 2006).

Moreover, designating a protected area is one thing, but establishing it *in situ* and managing it efficiently will depend on the financial and political supports of the local governments (Leverington et al. 2010; Pinto and Partidário 2012). As a result, designated PAs might suffer from a lack of adequate monitoring budget and trained staff (Hochkirch et al. 2013). The Natura 2000 Network has also recently been criticized for its lack of flexibility, adaptability, and monitoring (Crofts 2014; Hochkirch et al. 2013).

11.3 Agriculture and Conservation

Extensive agriculture is often associated with high biodiversity (EEA 2004; Halada et al. 2011). As a result, the concept of "High Nature Values Farmland" (HNVF) was introduced in the 1990s and now represents 15–25 % of the EU countryside (EEA 2004). High Nature Value Farmland areas typically depend on human activities, which maintain them by blocking the process of natural successions (EEA 2004; Halada et al. 2011; Merckx and Macdonald, in press; and see Chap. 6). In particular, some of the Natura 2000 sites are covered on more than a fourth of their area by extensive farmland (EEA 2004). In a review of the 231 habitats types of the Annex I of the Habitat Directive, Halada et al. (2011) identified 63 habitats depending on agricultural practices for their management, 23 of which are "fully dependent", while 40 "partly depend" on agriculture, mainly due to the prevention of natural successions.

High Nature Value Farmland areas are currently threatened by two opposing forces, intensification of agriculture on the one hand, and rural depopulation and farmland abandonment on the other hand (EEA 2004, 2009b). In 2003, the Kyiv Resolution on Biodiversity, made the identification and preservation of HNVF a conservation goal (EEA 2009b). This EU conservation strategy was later integrated into the second pillar of the CAP. Agri-environment schemes (AES) and other EU subsidies thus became a tool for High Nature Value Farmland conservation (EEA 2004).

Additionally, though the European Parliament recently stated that the EU biodiversity policies were not well integrated into other sectoral policies such as energy, transport, and agriculture (European Parliament 2009), agri-environmental policies have attempted for quite some times to better integrate agricultural productivity and biodiversity conservation. Currently, EU funds address the relationship between farmers and conservation in two ways. On the one hand, the EU compensates farmers receiving a lower income due to environmental restrictions. On the other hand, the EU created incentives for farmers to develop an environmentally friendly agriculture. Both forms of subsidies are not exclusive. Following the 2003 amendment of the regulation on Rural Development of the EU (1783/2003), farmers will receive monetary compensations for the "costs incurred and income foregone" resulting from the classification of their land as a Natura 2000 site according to Article 16(1). Articles 22–24 of the same regulation directly address AES, and how "support should be granted to farmers who give agri-environmental [...]

commitment for at least 5 years" (Article 23). The subsidies are destined to cover the "income foregone", "additional costs resulting from the commitment", and "the need to provide an incentive" (Article 24). The payment of subsidies and the implementation of agri-environmental policies vary greatly from one Member State to the other (EEA 2009b).

Nonetheless, the consequences of subsidizing nature conservation through the Common Agricultural Policy are debatable. First, a contradiction can emerge when the first pillar of the CAP favors intensive and productive agriculture on one hand, and hence fragments natural habitats (Crofts 2014; EEA 2009b; Henle et al. 2008), while, on the other hand, the second pillar incents farmers to develop environmentally friendly practices. Additionally, the compensations paid to farmers in Least Favored Areas (supported by the second pillar of the CAP to limit farmland abandonment) poses no real limits to intensification and overgrazing, provided that farmers follow country-specific "good farming practices" (EEA 2004). There is also no direct link between the amount spend in CAP subsidies per ha and the level of High Nature Value Farmland in an area (EEA 2004, 2009b; Halada et al. 2011). Finally, the payments of CAP subsidies in remote and less productive areas can appear inadequate so far. The phenomenon of rural depopulation was initiated in the 1950s in Western Europe, driven by socio-economic factors interacting to create a "circle of decline" in those remote areas (MacDonald et al. 2000; Rey Benayas et al. 2007), which is not likely to be interrupted, despite the rural development policies that have been implemented, and the resulting payment of subsidies (see Fig. 1.3 in Chap. 1).

The direct consequence of the phenomenon of rural depopulation is the abandonment of farmland in the less productive areas of the EU (see Chap. 1). Agricultural land abandonment is typically perceived negatively in developed countries (Meijaard and Sheil 2011; Queiroz et al. 2014), as a result of, *inter alia*, observed land encroachment, increased risk of fires, and decreases in populations of farmland birds. Yet, the withdrawal of human activities from those areas is also an (often disregarded) opportunity to increase the area of wilderness in the EU by applying rewilding as a land management policy.

11.4 Opportunities for Wilderness and Rewilding

Wilderness is both an ecological and social concept. The ecological meaning and extent of wilderness is multi-dimensional and varies depending on the metrics used (see Chap. 2). Wilderness is typically associated with the (quasi) absence of human impact, the large size of the area (e.g. 10,000 ha), and the naturalness of the dynamics that govern ecosystems (European Commission 2013; Fisher et al. 2010). The social and subjective concept of wilderness and wildlands is, for example, associated with the notions of remoteness and solitude (Fisher et al. 2010; Fritz et al. 2000). As a result, the definition of wilderness by the various people experiencing it will also depend on their perceptions of such areas (Nash 1967). Probably one of the most well known definition was given by the Wilderness Act of 1964 in the United States as " […] an area where the earth and its community of life are untrammeled by man, where man himself is a visitor who does not remain." (US Congress 1964). An area characterized as "wilderness" will thus be managed by "no-intervention" or "set-aside" practices (European Commission 2013). Europe is one of the continents with the least amount of wilderness (Mittermeier et al. 2003), mainly due to its long history of human induced land-use changes (see Chap. 8). Currently, the wilderness of the EU28 is mainly located in Scandinavia and in mountainous areas (see Chaps. 2 and 3).

Globally, wilderness and protected areas do not necessarily coincide. Though some wilderness areas might not currently require protection (e.g. due to their remoteness), wilderness conservation is a proactive measure that could pay off in the near future (Brooks et al. 2006). Using human density, size of the area, and historical intactness as metrics, Mittermeier et al. (2003) found that only 7% of the world's remaining wilderness was included in Protected Areas of IUCN categories I to IV. When looking into all types of Protected Areas in Europe, there is little to no correlation between the location of Nationally Designated Protected Areas and Natura 2000 areas with higher values in the Wilderness Quality Index (Fisher et al. 2010). However, there is a correlation between the occurrence of areas under the IUCN Categories I and II and high wilderness quality (Fisher et al. 2010). The number of Nationally Designated Protected Areas in Europe that falls under the IUCN Ib category ("wilderness areas") are located mainly in Sweden, Estonia, Slovakia, and Slovenia. Only 12 of the 28 EU Member States manage PAs designed in categories Ia or Ib, with different national legislation regarding the designation, the size of the area, the type of management and the level of human activity allowed (European Commission 2013).

Nevertheless, European wilderness is progressively gaining more importance, both in science, in conservation policy and at their interface. Its role in halting biodiversity loss was officially recognized (European Parliament 2009; Jones-Walters and Čivić 2010), with a will to include wilderness in the post–2010 targets. As a result, the European Parliament called for an effort to define both wilderness and the benefits derived from it, and for a better integration of wilderness in conservation policies. A special attention was to be given to wilderness areas within the Natura 2000 network. Indeed, some conceptual conflicts can arise when the noninterventional management of wilderness areas goes against the management of secondary (semi-natural) habitats of Annex I, such as the "European dry heaths" and "Dehesas with evergreen *Quercus spp.*" (Halada et al. 2011), unlike primary habitats, which rely on natural processes, for example "Western Taiga" and "Bog woodlands" (European Commission 2013; Fisher et al. 2010).

The European Commission (2013) recently published guidelines on the management of wilderness areas within the Natura 2000 Network. Though not legally binding for the Member States, they illustrate the will to include wilderness in EU conservation policies. The guidelines state that management practices for wilderness in the Natura 2000 network can involve the total or partial interdiction of human activities. When applicable, zonation can be used to define an area of non-intervention management for the wilderness core habitat, and a managed zone for secondary habitats. The guidelines also emphasize the importance of addressing local communities, to explain them the functioning of non-intervention management, and the benefits they could derive from it. Finally, scale has its importance in the designation and management of wilderness areas, as too little, or too fragmented land would not meet the criteria to allow for natural processes (European Commission 2013).

With the ongoing trends of farmland abandonment occurring on the continent, and the momentum gained by wilderness, rewilding appears as a valid land management option (see Chap. 1). It consists in the restoration of ecological processes and self-sustaining ecosystems, either passively, or with low to mild levels of intervention early on if the land-use history requires it (see Chaps. 7 and 8). Rewilding has proven to be beneficial to both biodiversity and human well-being (see Chaps. 1 and 3).

Increasing the area of wild land via the rewilding of abandoned landscapes will contribute to delineating new wilderness areas in the European landscape, with adequate conservation status and appropriate management. As such, rewilding can further increase the ecological coherence and connectivity of the protected areas in the EU28. Increasing the area of wilderness via rewilding will also contribute to the large scale natural processes that maintain it (e.g. European Commission 2013).

Some of the most emblematic species benefiting from land abandonment and rewilding are large mammals (Deinet et al. 2013; Enserink and Vogel 2006; Russo 2006; see Chaps. 1 and 2). They demand a large availability of land in order to sustain their dispersal and home range establishment requirements (Jones-Walters and Čivić 2010), and limit conflicts with humans, which also makes wilderness areas essential to their conservation (Mittermeier et al. 2003). Additionally, species listed in the Birds Directive, which are specialists of old-growth forests (e.g. the three-toed woodpecker—*Picoides tridactylus*), or which have large habitat requirements (e.g. the Siberian tit—*Parus cinctus*, the black woodpecker—*Dryocopus martius*), can benefit from the increase of wilderness areas (European Commission 2013).

The notion of a "perceived wilderness" (Jones-Walters and Čivić 2010) is important when investigating the benefits supplied by rewilded areas for people. For example, the increase in wild areas and the resulting wildlife comeback are thought to contribute to reconnecting Europeans with nature (Deinet et al. 2013). The cultural services provided via the enjoyment and experiencing of wilderness, for example the perception of solitude and remoteness, can reciprocally motivate its conservation and guide policies and land management. Wilderness areas also supply a wide range of provisioning and regulating services, such as freshwater provision, carbon sequestration, and nitrogen regulation (see Chap. 3).

Having in mind the potential benefits of rewilding and increased areas of wilderness, we can now investigate which could be their contribution to global and European conservation targets.

11.5   Global and European Conservation Targets

After failing to meet the biodiversity targets which had been set for 2010 (Butchart et al. 2010), the parties of the Convention on Biological Diversity (CBD) adopted an agreement in Nagoya, which set 20 Aichi Targets to preserve biodiversity and eco-system services by 2020 (CBD 2011). Several targets can be addressed by protected areas, wilderness, and rewilding. In particular, Target 11 requires that "at least 17 % of terrestrial and inland water […] are conserved through effectively and equitably managed, ecologically representative and well connected systems of protected areas […]". For most European countries, this target has already been reached, in the sense that most countries have more than 17 % of their national territory within a protected area (Fig. 11.2b), although effective management and wilderness conser-vation might fall short (e.g. Fig. 11.2c). For other targets, the level of completion is not so easily measured. Target 15 calls for the enhancement of ecosystems' resil-ience including through the "restoration of at least 15 % of degraded ecosystems", and the increase of carbon stocks. Rewilding is a particular case of restoration, and can contribute to the achievement of this target, particularly when looking into the increases in carbon stocks that could result from an enlargement of wild areas (see Chap. 3). Furthermore, Target 12 requires the prevention of the extinction of threat-ened species and the improvement of their conservation status. Again, the rewilding of abandoned landscapes, and an increase in wilderness areas, can directly contribute to this target, as several species already show increasing trends (Deinet et al. 2013; LCIE 2004; and see Chaps. 1 and 4). Finally, Target 7 requires that "areas under ag-riculture, aquaculture and forestry are managed sustainably, ensuring conservation of biodiversity", while Target 3 calls for the termination, or the reform, of "incen-tives, including subsidies, harmful to biodiversity". Both these tasks could be ad-dressed by a reform of the subsidies system of the CAP and the AES, and their shift towards rewilding and the restoration of wild lands in low income agricultural areas (e.g. Merckx and Pereira, in press).

The Aichi Targets and their implications are not legally binding for countries. Nonetheless, the EU and all its Member States adopted the conservation targets in the European Biodiversity Strategy and defined a new regional strategy to 2020 (Table 11.2), in order to both halt biodiversity loss and restore degraded systems (European Commission 2011; Hochkirch et al. 2013). Some of these targets can be addressed by an efficient, and when needed better designed, network of PAs. The preservation of wilderness and the increase in wild areas is also considered as playing a crucial role in reaching some of the targets (European Commission 2013), namely "protecting and restoring biodiversity and ecosystem services" (Targets 1 and 2), and "reducing pressures on biodiversity" (Targets 3 and 5). Additionally, wilderness areas, being remote and not densely populated, present the advantage of lower land prices per hectare, while non-intervention implies drastically lower management costs (Mittermeier et al. 2003).

Table 11.2 EU targets and biodiversity strategies to 2020, most relevant within the context of protected areas, wilderness and rewilding discussed in this chapter. (European Commission 2011)

European targets	Status in 2010	Objective for 2020
1. Implement the habitat and bird directives	17% of habitats and species protected by the Habitats directive are in favorable status	34% of the habitats and 26% of the species should either improve or be in a favorable status
	52% of the bird species are in a secure position	80% of bird species should be secured or improving
2. Maintain and restore ecosystems and their services	No continental data on degraded ecosystems, and the supply of ecosystem services	Increase knowledge and define actions • Mapping and assessment of the state of ecosystems and their services • Definition of a strategic restoration framework, including with the development of green infrastructures • Ensure no let loss of biodiversity and ecosystem services
3. Increase the contribution of agriculture and forestry to biodiversity	Only 15–25% of extensive high nature value farmland remains	Maximize agricultural areas covered by biodiversity measures of the CAP • Enhance direct payments for environmental public goods in the EU CAP • Better target Rural development to biodiversity conservation. • Conserve Europe's agricultural genetic diversity
	7% of the habitats and 3% of the species protected by the Habitats Directive and depending on agriculture have a favorable status	
	Farmland bird populations have decreased by 50% since 1980 but have now leveled of	
	Farmland butterfly populations have decreased by 70% since 1990	
	21% of forest habitats and 15% of forest species protected under the habitat directive have a favorable status	Forest management plans, in line with sustainable forests management are in place for all publicly owned forest and forest holdings above a certain size • Encourage forest holders to protect and enhance forest biodiversity • Integrate biodiversity measures in forest management plans
	1–3% of forests are in natural and unmanaged status	

The EU incorporated the Aichi Target 3 to its plan, in particular to "reform, phase out and eliminate harmful subsidies at both EU and Member States level" (Target 6–Action 17c). At the same time, the Commission highlights the importance of integrating biodiversity policies into wider European policies concerns such as agriculture and forestry, and to "minimize the duplication of effort and maximize synergies between efforts undertaken at different levels" (European Commission

2011). In a context of farmland abandonment in remote and less productive areas, maximizing the synergies between conservation efforts can be done by redirecting subsidies towards rewilding (Merckx and Pereira, in press, and see Chap. 6), while allowing the remaining local population to live off the land through different means than its cultivation. Moreover, an efficient implementation of rewilding for the management of the abandoned land will have, in the long run, a positive impact on biodiversity and the supply of ecosystem services (see Chaps. 1, 3). The latter includes cultural services, such as ecotourism, which will directly benefit local populations.

11.6 Recommendations for Rewilding

The current European policy response to pressures on biodiversity can be either with site protection (e.g. SPAs SACs), or with the regulation of the activities of those exploiting the land, which can also be relying on voluntary actions, i.e. with Agri-Environmental Schemes (EEA 2004). Rewilding abandoned farmland can efficiently contribute to reaching European and global conservation targets. But in order to do so, a policy framework must be designed to include rewilding in the land management options given to practitioners (see Chap. 1). To that extent, European conservation policies must aim toward several goals.

In places where people still keep a strong link with nature, a wilderness comeback via natural regeneration should not be excessively problematic (McGrory Klyza 2001). Yet, when the link with traditional landscapes is the strongest, as in many regions of Europe, rewilding might be perceived negatively (Bauer et al. 2009; Hochtl et al. 2005). Communication between scientists, policy-makers, decision-makers, and the public will be essential to allow the implementation of rewilding, and to promote the values of wilderness in a landscape. Development initiatives are also known to ease the transitions between one form of management and another, for instance by increasing the support of local communities for the protected area (Pinto and Partidário 2012). Giving the opportunities to populations to shift their activities from low-income agriculture to ecotourism in rewilded areas can be an efficient way to meet both ends (see Chaps. 3 and 9).

The proposed "greening" reform of the CAP could further compensate stakeholders maintaining low productive practices in order to preserve traditional agricultural habitats (Hochkirch et al. 2013). Another option is to maintain payments for farmers that apply environmentally friendly practices on productive soils, and redirect subsidies on less productive lands towards rewilding (Merckx and Pereira, in press). By doing so, Member States will still be able to meet the demands for agricultural goods, yet promoting responsible and green practices on productive soils, while the lands left abandoned due to their remoteness, their lower productivity, and the difficulty to cultivate them (MacDonald et al. 2000; Rey Benayas et al. 2007, and see Chap. 1) will be rewilded and managed for other activities linked with wilderness. Such approach can be seen as land-sharing at the local scale (with environmentally friendly agriculture), while at a broader scale food production and wilderness will occur on different areas, i.e. land-sparing (Merckx and Pereira, in press; Phalan et al. 2011).

When a transition from "species conservation" to "species management" occurs, adapted policy tools will be needed (Henle et al. 2013). Some of the species benefiting from rewilding and showing positive population trends with land abandonment are large mammals, which are often associated with human/wildlife conflicts (see Chap. 1). If those populations were to increase substantially, it could be difficult to segregate them entirely to wilderness areas and mechanisms will have to be designed to allow for mitigation, compensation and/or cohabitation (e.g. large carnivores–see Chap. 4, and large scavengers–see Chap. 5). The set of policy instruments that can address human/wildlife conflicts are: regulatory (i.e. referring to the management and control of species); economic (e.g. compensations for damages caused by wildlife, subsidies for technical development for the prevention of damages); and educational, directed at the civil society (Similä et al. 2013).

Promoting rewilding to manage abandoned farmland means shifting the policies towards an ecosystem process-based conservation, rather than the static conservation of a set of species and habitats which is the current paradigm (Hochkirch et al. 2013). Assisted restoration can be needed in the early stages of conservation, depending on the ecological filters that could prevent and/or limit the return to self-sustaining ecosystems (see Chaps. 1, 7, and 8). For instance, the restoration of disturbance regimes to rewild opened landscapes following the abandonment of pastoral activities will mean the need of wild, or semi-wild grazers (see Chap. 8), which could be (re)introduced if no local population was present. Though the introduction of wild species is legally framed (IUCN 2013), it is not the case for domestic species, such as horses, which could be used to maintain the disturbance regime of abandoned pastures. This calls for a legal framework on their reintroductions and on the liability of the various stakeholders involved (Jones-Walters & Čivić 2010).

Rewilding will help policy-makers and stakeholders in rethinking their relationship with nature. In particular, the opportunity given by farmland abandonment to passively restore millions of hectares of land could give Europe an occasion to end the trends of double-standards between developed and developing countries in regard to conservation policies. For instance, deforestation is (rightfully) considered as a major degradation of ecosystems in developing countries, yet EU countries subsidies the maintenance of low productive agriculture to limit secondary successions on their land (Meijaard and Sheil 2011). Rewilding thus needs to gain visibility in the public and political sphere, as saliency (e.g. mainstreaming the concept of rewilding) has proven to be essential to the integration of concepts and ideas into the policy agendas (Jørgensen et al. 2014; Rudd 2011). In particular, rewilding research should aim at having three important impacts on policy makers (Rudd 2011): a conceptual impact (to change the way policy makers think), an instrumental impact (to directly influence existing policies and managements), and a symbolic impact (to support established positions).

Changes in what societies want to preserve, and how they protect it have already been observed (e.g. Pinto and Partidário 2012). The conservation and management of the European biodiversity has evolved since the 1970s (Fig. 11.1), giving for instance increasing importance to the role of local communities in managing Protected Areas, and to the benefits that they should get from those (Jones-Walters and Čivić 2013). For better or for worse, throughout decades of transitions in the way biodiversity is preserved, conservation baselines shifted, decision makers and stakeholders adapted, and so did the management approaches. Bringing rewilding in the agenda of conservation policies by showing its potential to both tackle the issue of land abandonment and restore wilderness could lead the way to a new transition of biodiversity conservation in Europe.

Acknowledgments We thank Jörg Freyhof, Silvia Ceauşu and Alexandra Marques for discussions and comments on earlier versions of the manuscript. L.M.N. was supported by a PhD fellowship from the FCT (SFRH/BD/62547/2009).

References

Bauer, N., Wallner, A., & Hunziker, M. (2009). The change of European landscapes: Human-nature relationships, public attitudes towards rewilding, and the implications for landscape management in Switzerland. *Journal of environmental management, 90,* 2910–2920.

Borrini-Feyerabend, G., Dudley, N., Jaeger, T., Lassen, B., Pathak Broome, N., Phillips, A., & Sandwith, T. (2013). *Governance of protected areas. From understanding to action.* Switzerland: IUCN.

Brooks, T. M., Mittermeier, R. A., da Fonseca, G. A., Gerlach, J., Hoffmann, M., Lamoreux, J. F., Mittermeier, C. G., Pilgrim, J. D., & Rodrigues, A. S. (2006). Global biodiversity conservation priorities. *Science, 313,* 58.

Butchart, S. H. M., Walpole, M., Collen, B., van Strien, A., Scharlemann, J. P. W., Almond, R. E. A., Baillie, J. E. M., Bomhard, B., Brown, C., Bruno, J., et al. (2010). Global biodiversity: Indicators of recent declines. *Science, 328,* 1164–1168.

CBD (2011). *Aichi biodiversity targets.*

Coetzer, K. L., Witkowski, E. T. F., & Erasmus, B. F. N. (2014). Reviewing biosphere reserves globally: Effective conservation action or bureaucratic label? *Biological Review, 89,* 82–104.

Crofts, R. (2014). The european Natura 2000 protected area approach: A practitioner's perspective. *Parks 20,* 75–86.

Deinet, S., Ieronymidou, C., McRae, L., Burfield, I. J., Foppen, R. P., Collen, B., & Bohm, M. (2013). *Wildlife comeback in Europe: The recovery of selected mammal and bird species.* London: Final report to Rewilding Europe by ZSL, BirdLife International and the European Bird Census Council.).

Dudley, N. (2008). *Guidelines for applying protected area management categories.* Gland, Switzerland: IUCN.

EEA. (2004). *High nature value farmland: Characteristics, trends and policy challenges.* Copenhagen: European Environment Agency.

EEA. (2009a). *State of progress by Member States in designating sufficient protected areas to provide for Habitats Directive (92/43/EEC) Annex I habitats and Annex II species.* Copenhagen: European Environment Agency.

EEA. (2009b). *Distribution and targeting of the CAP budget from a biodiversity perspective.* Copenhagen: European Environment Agency.

EEA. (2013a). *Nationally designated areas (CDDA—1).* Accessed on line April 2014.

EEA. (2013b). *Natura 2000 data-the European network of protected sites*. Accessed on line April 2014.

Enserink, M., & Vogel, G. (2006). The carnivore comeback. *Science, 314,* 746.

European Commission. (2002). *Commission Working Document on Natura 2000* (Brussels).

European Commission. (2011). *The EU Biodiversity Strategy to 2020* (Luxembourg).

European Commission. (2013). *Guidelines on Wilderness in Natura 2000. Management of wilderness and wild areas within the Natura 2000 Network*. Technical Report - 2013–069.

European Parliament. (2009). *European Parliament resolution of 3 February 2009 on Wilderness in Europe* (2008/2210(INI)).

Fisher, M., Carver, S., Kun, Z., McMorran, R., Arrell, K., & Mitchell, G. (2010). *Review of status and conservation of wild land in Europe*. (Project commissioned by the Scottish Government.).

Fritz, S., Carver, S., & See, L. (2000). New GIS approaches to wild land mapping in Europe. In *Wilderness Science in a Time of Change Conference*.

Gaston, K. J., Jackson, S. F., Nagy, A., Cantú-Salazar, L., & Johnson, M. (2008a). Protected areas in Europe. *Annals of the New York Academy of Sciences, 1134,* 97–119.

Gaston, K. J., Jackson, S. F., Cantú-Salazar, L., & Cruz-Piñón, G. (2008b). The ecological performance of protected areas. *Annual Review Ecology, Evolution and Systematics, 39,* 93–113.

Halada, L., Evans, D., Romão, C., & Petersen, J.-E. (2011). Which habitats of European importance depend on agricultural practices? *Biodiversity and Conservation, 20,* 2365–2378.

Henle, K., Alard, D., Clitherow, J., Cobb, P., Firbank, L., Kull, T., McCracken, D., Moritz, R. F. A., Niemelä, J., Rebane, M., et al. (2008). Identifying and managing the conflicts between agriculture and biodiversity conservation in Europe—A review. *Agricultural Ecosystem Environment, 124,* 60–71.

Henle, K., Kranz, A., Klenke, R. A., & Ring, I. (2013). Policy brief. In R. A. Klenke, I. Ring, A. Kranz, N. Jepsen, F. Rauschmayer, & K. Henle (Eds.), *Human-wildlife conflicts in Europe* (pp. 1–3). Berlin: Springer.

Hochkirch, A., Schmitt, T., Beninde, J., Hiery, M., Kinitz, T., Kirschey, J., Matenaar, D., Rohde, K., Stoefen, A., Wagner, N., et al. (2013). Europe needs a new vision for a natura 2020 network. *Conserv. Lett, 6,* 462–467.

Hochtl, F., Lehringer, S., & Konold, W. (2005). "Wilderness": What it means when it becomes a reality-a case study from the southwestern Alps Landscape. *URBAN Plan, 70,* 85–95.

IUCN. (1969). *Tenth General Assembly-Volume II: Proceedings and Summary of Business*. (Morges, Switzerland).

IUCN. (2013). *Guidelines for reintroductions and other conservation translocations*. (Version 1.0.). Switzerland: IUCN Species Survival Commission.

Jones-Walters, L., & Čivić, K. (2010). Wilderness and biodiversity. *Journal for Nature Conservation, 18,* 338–339.

Jones-Walters, L., & Čivić, K. (2013). European protected areas: Past, present and future. *Journal for Nature Conservation, 21,* 122–124.

Jørgensen, D., Nilsson, C., Hof, A. R., Hasselquist, E. M., Baker, S., Chapin, F. S., Eckerberg, K., Hjältén, J., Polvi, L., & Meyerson, L. A. (2014). Policy language in restoration ecology. *Restoration Ecology, 22,* 1–4.

LCIE. (2004). *Status and trends for large carnivores in Europe* (UNEP-WCMC Project).

Leverington, F., Costa, K. L., Pavese, H., Lisle, A., & Hockings, M. (2010). A global analysis of protected area management effectiveness. *Environmental Management, 46,* 685–698.

MacDonald, D., Crabtree, J. R., Wiesinger, G., Dax, T., Stamou, N., Fleury, P., Gutierrez Lazpita, J., & Gibon, A. (2000). Agricultural abandonment in mountain areas of Europe: Environmental consequences and policy response. *Journal of environmental management, 59,* 47–69.

McGrory Klyza, C. (2001). An eastern turn for wilderness. In C. McGrory Klyza (Ed.), *Wilderness comes home. Rewilding the Northeast* (pp. 3–26). London: Middlebury College Press.

Meijaard, E., & Sheil, D. (2011). A modest proposal for wealthy countries to reforest their land for the common good. *Biotropica, 43*(5), 524–528.

Merckx, T., & Macdonald, D. W. (in press). Landscape-scale Conservation of Farmland Moths. In D. W. Macdonald & R. E. Feber (Eds.), *Wildlife conservation on farmland*. Oxford University Press.

Merckx, T., & Pereira, H. M. (in press). Reshaping agri-environmental subsidies: From marginal farming to large-scale rewilding. *Basic and Applied Ecology*. doi:10.1016/j.baae.2014.12.003

Mittermeier, R. A., Mittermeier, C. G., Brooks, T. M., Pilgrim, J. D., Konstant, W. R., Da Fonseca, G. A. B., & Kormos, C. (2003). Wilderness and biodiversity conservation. *Proceedings of the National Academy of Science, 100,* 10309–10313.

Nash, R. (1967). *Wilderness and the American mind.* New Haven Yale: Yale University Press.

Phalan, B., Onial, M., Balmford, A., & Green, R. E. (2011). Reconciling food production and biodiversity conservation: Land sharing and land sparing compared. *Science, 333,* 1289–1291.

Pinto, B., & Partidário, M. (2012). The history of the establishment and management Philosophies of the Portuguese protected areas: Combining written records and oral history. *Environmental Management, 49,* 788–801.

Possingham, H., Wilson, K. A., Andelman, S. J., & Vynne, C. H. (2006). Protected areas: Goals, limitations, and design. In M. J. Groom, G. K. Meffe, and C. R. Carroll (Eds.), *Principles of Conservation Biology* (pp. 507–549). USA: Sinauer Associates, Inc.

Queiroz, C., Beilin, R., Folke, C., & Lindborg, R. (2014). Farmland abandonment: threat or opportunity for biodiversity conservation? A global review. *Frontiers in Ecology and Environment, 12*(5), 288.

Ramão, C., Reker, J., Richard, D., & Jones-Walters, L. (2012). *Protected areas in Europe-an overview.* Copenhagen: European Environment Agency.

Rey Benayas, J. M., Martins, A., Nicolau, J. M., & Schulz, J. J. (2007). Abandonment of agricultural land: an overview of drivers and consequences. *CAB reviews: Perspectives in agriculture, veterinary science, nutrition and natural resources, 2,* 1–14.

Rudd, M. A. (2011). How research-prioritization exercises affect conservation policy. *Conservation biology: the journal of the Society for Conservation Biology, 25,* 860–866.

Russo, D. (2006). *Effects of land abandonment on animal species in Europe: conservation and management implications.* University Degli Studi Napoli Federico Napoli Italy.

Similä, J., Varjopuro, R., Habighorst, R., & Ring, I. (2013). Module 4: Legal and institutional framework. In R.A. Klenke, I. Ring, A. Kranz, N. Jepsen, F. Rauschmayer, & K. Henle (Eds.), *Human-Wildlife Conflicts in Europe* (pp. 251–260). Berlin: Springer.

UNESCO. (1996). *Biospheres reserves: The seville strategy and the statutory framework of the world network.* Paris: UNESCO.

US Congress. (1964). Wilderness Act—Public Law 88-577 (16 U.S.C. 1131–1136) Washington, DC.

World Heritage Centre. (2013). *Operational guidelines for the implementation of the world heritage convention.* Paris: UNESCO.

Index

A
Agri-environment schemes (AES), 114, 137, 213, 217
Air quality, 48, 49, 52, 57

B
Baseline, 12, 27, 145, 162, 211
Bear, 15, 68, 69, 72, 138, 156
Birds of prey, 12, 28, 36, 86, 120, 135, 136
Butterflies, 108, 109, 115, 118, 120, 121

C
Carnivore recovery, 36
Carrion pulsed resource, 87, 96, 98
Climate regulation, 48, 57
Climax, 111, 112
CLUE model, 11, 18, 37, 53, 98
    Dyna-CLUE model, 38
Coexistence, 68, 80, 81, 89
Compensations, 213, 214, 220
Competences, 193, 194, 197, 198, 201–203
    business, 198, 200, 202
    complexity, 198, 199, 202
    normative, 198, 199, 202
    opportunity, 198, 202
    social, 198, 199, 202

D
Designing curricula, 192
Diversity
    alpha, 118, 145, 155, 156
    beta, 117, 118, 145, 156
    gamma, 156

E
Ecological
    resilience, 15, 16

restoration, 58, 128, 129, 137, 139, 163, 196
Economic benefits, 52, 58–61, 192
Ecosystem engineers, 159
    pre-neolithic, 148, 149
Ecosystem services, 5–7, 12, 17, 27, 48, 49, 53, 57, 60
Ecotourism, 56, 99, 101, 172, 219
Education, 8, 137, 192–194, 196, 197, 202
Entrepreneurship, 192, 193
Erasmus Intensive Programme, 194
EU conservation policies, 215
European
    landscape, 79, 86, 87, 91, 100, 145, 146
    legislation, 69
    Rewilding Network, 173
    Safari Company, 180, 183, 189
    Wildlife Bank, 176
Extinctions, 75, 111, 148, 149, 156

F
Farmland
    abandonment, 38, 39
    effects on macro-moths, 115, 116, 118
Fire, 14, 39, 116
    -dependent ecosystem, 153, 154
    regime, 149, 153–155, 160, 162
    suppression, 39, 144, 154
Food webs, 86

G
Guilds, 89, 94, 98

H
Habitat, 118, 159, 210
    heterogeneity, 110, 112, 118–121, 158
    restoration, 129, 137

H. M. Pereira, L. M. Navarro (eds.), *Rewilding European Landscapes,*
DOI 10.1007/978-3-319-12039-3, © The Author(s) 2015